I0050276

TIME BETTERING DAYS
and Other Essays

TIME BETTERING DAYS
and Other Essays

by
EUGEN ROSENSTOCK-HUESSY
foreword by Knut Martin Stünkel

WIPF & STOCK · Eugene, Oregon

TIME BETTERING DAYS AND OTHER ESSAYS

Copyright © 2025 Norman Fiering. All rights reserved. Except for brief quotations in critical publications or reviews, no part of this book may be reproduced in any manner without prior written permission from the publisher. Write: Permissions, Wipf and Stock Publishers, 199 W. 8th Ave., Suite 3, Eugene, OR 97401.

Wipf & Stock
An Imprint of Wipf and Stock Publishers
199 W. 8th Ave., Suite 3
Eugene, OR 97401

www.wipfandstock.com

PAPERBACK ISBN: 979-8-3852-1453-2
HARDCOVER ISBN: 979-8-3852-1454-9
EBOOK ISBN: 979-8-3852-1455-6

05/16/25

Previously published by Argo Books, 1981, as *Rosenstock-Huessy Papers*, volume 1

In memory of Clinton C. Gardner,
who kept the faith, brilliantly

CONTENTS

FOREWORD

WHY SHOULD WE LISTEN to Eugen Rosenstock-Huessy these days? One of the particularly encouraging aspects of Rosenstock-Huessy's thinking is that he strongly urges people to say what is currently on their mind. For him, this kind of speaking is not a deficient or unfinished form of communication. Only in this unguarded speaking, not in prefabricated statements and opinions, he teaches, can something significant happen for the future of the human race.

Today's intellectual climate is characterized by a refusal to communicate. While the possibilities for communication seem to be rapidly increasing, the willingness to communicate is decreasing to the same degree. And more than that: *not* talking to someone who takes an opposing position to one's own is increasingly perceived as a virtue and is almost emphatically propagated. People today experience that they can expect praise if they refuse to communicate, engage in dialogue, or talk to others.

It is not possible to attribute blame for this situation unilaterally. Progressives and conservatives, left and right, cultural, religious and social groups indulge with growing enthusiasm in a disastrous cult of their own and never tire of discrediting opposing positions, not least morally: the "other" is that which is actually impossible. The others are denied the ability to speak to the opposing group at all: The "West" it is claimed, cannot understand the "East" nor can the "North" understand the "South," and *vice versa*; every attempt to do so is perceived to be an encroachment and a ravishment of one's own. The result of this idea is a growing division within what is called "society"; a segmentation of groups that develop rigid internal demarcation criteria towards the outside and enforce them in practice: in politics, in sociality, in religion and also in science and scholarship. Such demarcated groups make a seductive promise to the individual: they promise security, above all self-security, unchallenged by contradiction. One finds oneself in and argues only according to the criteria of one's own bubble.

Eugen Rosenstock-Huessy, on the other hand, is someone with whom readers and listeners can and should enter into a dialogue. He has interesting things to say and can show us how to look at familiar things from a new and unusual angle. Rosenstock's works offer the reader a multitude of possible "aha" experiences, be it in the field of pedagogy, andragogy, historiography, in the philosophy of language or the understanding of space and time. Above all, however, he can turn a dialogue into a conversation by provoking contradiction. For a successful conversation, it is by no means necessary for the speakers to be in agreement, neither at the beginning nor at the end of the conversation. This is why no one is as unsuitable as this author to be the master of a devoted following. If one remains faithful to him and his ideas, especially his ideas about language, one contradicts them, becomes a contradictor of them—just as Rosenstock himself becomes a contradictor of his own unquestioned ideas and convictions. In contradicting, the respective "*Widerwart*" bears witness to the common future. The notion of *Widerwart*, literally someone who wards or guards against something with the tendency to become annoying to his opponent, was introduced in the discussion between Rosenstock and his friend Franz Rosenzweig on the Jewish-Christian relationship, and denotes a friend in contraposition, i.e., someone who is willing to articulate and defend his opposing point of view persistently.

Through his insistence on the creativity of dialogical contradiction, Rosenstock is a genuinely sociological author, namely an author against the absolutization of his own bubble. He believes in plurality and the intertwining of different time rhythms and languages. What can one do against self-segregating bubbles? You have to make them bubble over. In his later work, Rosenstock coined a word for this: *symblysma*—the overflowing of bubbles.

A future-oriented result emerges from the conversation with *Widerwarten*. According to Rosenstock, it may be that a situation arises in and through the conversation, in which a certain group responds individually in a certain way, which he describes as symblysma, a term that appears in a path-breaking section of his book "The Breath of the Spirit" (*Der Atem des Geistes*). This is characterized by the "common awakening of the spirits," which is expressed in the fact that "many voices become loud because one spirit overflows." In this way, one of Rosenstock's demands on sociology is fulfilled, namely that "people who have different birth certificates bear witness together to the common future." With the symblysma, a "new basic category that sociology needs," namely that of reciprocity, is gained. In his relevant essay "Symblysma or the exuberance of the Jesuits" (*Symblysma oder der Überschwang der Jesuiten*), Rosenstock attributes the discovery of this symblysma, for which he himself claims only to have found the word that is "short and striking enough to become naturalized," to six authors of Jesuit provenance, namely Peter Browe (who is unnamed in the text), then Peter Lippert, Hugo and Karl Rahner, Carlo Passaglia, and Hans Urs von Balthasar, whose writings probably had the most symblysmatic relevance for him and who coined a term corresponding to symblysma, namely that of *circumincessio*.

Translated as "common effervescence," symblysma is a catchy neologism to describe a well-known but not scientifically recognized process, namely the work of the Holy Spirit in the Pentecost event. Rosenstock sees his reflections on the symblysma as a contribution to biography: he is concerned with the hitherto neglected "history of 'biographical groups'," which can be presented as a "chapter in the history of the Holy Spirit." The work of the Holy Spirit goes hand in hand with a drastic biographical change in the lives of those affected. A group whose shared biography is characterized by a symblysmatic event, i.e. mutual enthusiasm, attains a special kind of insight, an inspiration in the form of responsible knowledge and responsible science, and a sense of the future.

Symblysma is thus also Rosenstock's answer to the loss of significance that Pentecost suffered as a group-forming and language-conferring inspiration in contemporary Christianity and which he repeatedly lamented. In a speech he claimed: "Pentecost has had bad times." In view of the overriding significance, also methodological, that Rosenstock ascribes to the Pentecost event, it is no coincidence that the name of his early friend Karl Barth and the desire for Pentecostal reconciliation appear in the context of such considerations. Rosenstock writes: "All worldly slogans of partisanship must perish in

the great readmission of Pentecost; for this is what true reason teaches." According to Rosenstock in his article "Pentecost and Mission,"
without the respective and continued recapitulation of Pentecost, the
memory of the original Pentecost "creates nothing but despair."

To Rosenstock, symblysma describes the polyphony of the Holy
Spirit. This is incompatible with the usual biographical categories that
are applied when different individuals meet intellectually, especially
with the contrast of one being right and correspondingly the other
being wrong, that biographers of individuals like to use. Speaking less
theologically and more sociologically: Rosenstock uses the term symblysma to describe the literally inspiring, biographically incisive effect
of the dynamic clashes within certain groups of different individuals.
At the same time, the notion takes the fact of Christianity seriously.
In his article "The Language of the West" (*Die Sprache des Westens*)
from his book "The Secret of the University" (*Das Geheimnis der Universität*) he claims:

"Not the solitary monk, not the solitary creative genius, but the
individual group that remains true to its individuality in the masses
will bear witness to the Holy Spirit: temporary and mortal like marriage and friendship, but also animated and spiritualized like all Pentecostal creatures."

In such a creaturely circle, truth happens, a truth that proves itself
practically and publicly through the responsibility, on the one hand,
of the respective individual towards the challenging event, but on the
other hand, also in the face of each other and, above all, in front of an
external audience. Truth becomes manifest in public speaking, which
explicitly includes speaking via print in books, publishing houses, and
magazines. In "Out of Revolution" it says: "Inspiration, genius, talents, or simply thought, language, writing, are not merely means for
starring lonely individuals; they are uniting people in a common life."
The publishing of books thus becomes a necessity of life, as a result of
an overwhelming symblysmatic event. His confessional text "I Am an
Impure Thinker" states:

"To write a book is no luxury. It is a means of survival. By writing
a book, a man frees his mind from an overwhelming impression. The
test for a book is its lack of arbitrariness, the fact that it had to be done
in order to clear the road for further life and work."

For Rosenstock, symblysma as a moving and productive event stands for a "unity of the spirit," which is also possible with non-contemporaries, i.e., a symblysma "between entire epochs". According to this claim, a symblysmatic experience is, for example, also possible between adherents of different religious traditions, without essentially endangering one side, because "the original Pentecost [was] not a phenomenon of conversion, but one of multilingual revelation. . . . " Symblysma is the expression of a linguistic unity that is more than the sum of the sentences or the sum of the speakers. In this respect, symblysma is a reformulation of Rosenstock's earlier insight that language is wiser than the one who speaks it.

In any case, the decisive factor in Rosenstock's concept of symblysma is therefore the establishment of a common spirit in a group of individuals, which, however, is essentially directed outwards and has an effect on the future and its possibilities of communication to others (corresponding to the Pentecost event). An outwardly closing spirit, a bubble, is unspiritual and not sustainable.

As an untimely contemporary (*Unzeitgenosse*), as a thinker from the beginning and middle of the last century, Rosenstock can have a lot to say to contemporaries of the 2020s. The following texts demonstrate this in an impressive way. Last but not least is his teaching that in situations of conversation there is hope for a symblysmatic event to shape a common, resistant but peaceful future. One only has to want to allow this situation of conversation—it would be "time bettering days."

*

WARUM SOLLTE MAN HEUTZUTAGE auf das Wort Eugen Rosenstock-Huessys hören? Es ist einer der besonders bestärkenden Aspekte des Denkens Rosenstock-Huessys, daß man von ihm nachdrücklich ermutigt wird, auszusprechen, was man gerade auf dem Herzen hat. Dabei ist für ihn diese Art des Sprechens keine defiziente Form der Kommunikation. Nur in diesem ungeschützten Sprechen, nicht in vorgefertigten Statements und Meinungen, so lehrt er, könne etwas Bedeutsames für die Zukunft des Menschengeschlechts geschehen.

Die geistige Situation der heutigen Zeit ist durch Kommunikationsverweigerung bestimmt. Während die Kommunikationsmöglichkeiten

immer größer zu werden scheinen, nimmt die Kommunikationsbe-
reitschaft immer mehr ab. Und mehr als das: Nicht mit jemandem zu
sprechen, der eine entgegengesetzte Position vertritt als die eigene,
wird zunehmend als Tugend empfunden und geradezu emphatisch
propagiert. Der Mensch von heute macht die Erfahrung, daß er mit
Lob rechnen kann, wenn er eine Kommunikation, einen Dialog oder
ein Gespräch mit anderen verweigert.

Eine einseitige Schuldzuschreibung an dieser Situation ist nicht
möglich. Progressive und Konservative, Linke und Rechte, kulturelle,
religiöse und soziale Gruppen frönen mit wachsender Begeisterung
einem verhängnisvollen Kult des Eigenen und werden nicht müde,
Gegenpositionen zu diskreditieren, nicht zuletzt auch moralisch: das
Andere ist das, was eigentlich unmöglich ist. Den Anderen wird die
Fähigkeit abgesprochen, mit der jeweiligen Gruppe überhaupt sprechen zu können: ‚der Westen‘, so wird behauptet, kann ‚den Osten‘
nicht verstehen, ebensowenig ‚der Norden‘, ‚den Süden‘; jeder Versuch
ist ein Übergriff und eine Vergewaltigung des Eigenen. Die Folge ist
eine wachsende Spaltung dessen, was man ‚die Gesellschaft‘ nennt;
eine Segmentierung von Gruppen, welche intern rigide Abgrenzungs-
kriterien nach außen entwickeln und praktisch durchsetzen: in der
Politik, in der Sozialität, in der Religion und auch in der Wissenschaft.
Solcherart abgegrenzte Gruppen machen dem Einzelnen ein verfüh-
rerisches Versprechen: sie versprechen Sicherheit, vor allem Selbst-
Sicherheit, unangefochten von Widerspruch. Man befindet sich in
und argumentiert nur gemäß der Kriterien der eigenen Blase.

Eugen Rosenstock-Huessy ist demgegenüber jemand, mit dem
man als Leser und Hörer in einen Dialog treten kann und auch in
Dialog treten soll. Er hat interessante Dinge zu sagen und kann zeigen,
wie man bekannte Dinge unter einem neuen ungewöhnlichen Blick-
winkel betrachtet. Rosenstocks Texte bieten dem Leser eine Vielzahl
von möglichen Aha-Erlebnissen, sei es im Bereich der Pädagogik, der
Andragogik, der Geschichtsschreibung, der Sprache, der Zeit. Vor allem aber kann er einen Dialog zu einem Gespräch machen, und zwar,
in dem er zum Widerspruch reizt. Für ein gelingendes Gespräch ist es
nämlich keinesfalls notwendig, daß die Sprechenden einer Meinung
sind, weder zu Anfang noch am Ende des Gesprächs. Daher ist auch
niemand so ungeeignet als Meister einer ergebenen Gefolgschaft wie
eben dieser Autor. Bleibt man ihm und seinen Ideen, zumal seinen

Ideen über die Sprache treu, widerspricht man ihnen, wird an ihnen zum Widerwart – ganz entsprechend wie Rosenstock selbst zum Widerwart der eigenen unhinterfragten Ideen und Überzeugungen wird. Im Widerspruch legt der jeweilige Widerwart von der gemeinsamen Zukunft Zeugnis ab.

Durch sein Insistieren auf die Kreativität des dialogischen Widerspruchs ist Rosenstock ein genuin soziologischer Autor, nämlich ein Autor gegen die Verabsolutierung der eigenen Blase. Er glaubt an Pluralität und das Ineinander verschiedener Zeitrhythmen und Sprachen. Was kann man gegen sich selbst abschottende Blasen tun? Man muß sie zum Überschäumen bringen. Rosenstock hat hierfür ein Wort geprägt: Symblysma – das Überfließen einer Blase.

Aus dem Gespräch mit Widerwarten entsteht ein zukunftsfähiges Ergebnis. Es kann nämlich, so Rosenstock, sein, daß sich im und durch das Gespräch eine Situation ergibt, in der eine bestimmte Gruppe individuell verbunden auf eine bestimmte Weise antwortet, die er mit einem später, in einem Text seines Buches *Der Atem des Geistes* geprägten Ausdruck als ‚Symblysma‘ bezeichnet. Dieses ist gekennzeichnet durch den „gemeinsamen Aufbruch der Geister", der sich darin äußert, „daß viele Stimmen laut werden, weil *ein* Geist überströmt." Auf diese Weise wird eine Anforderung Rosenstocks an die Soziologie erfüllt, nämlich, daß „Leute, die verschiedene Geburtsatteste haben, zusammen Zeugnis ablegen von der gemeinsamen Zukunft." Mit dem Symblysma wird eine „neue Grundkategorie, welche die Soziologie braucht", nämlich die der Gegenseitigkeit, gewonnen. Die Entdeckung dieses Symblysmas schreibt Rosenstock, für das er selbst nur das Wort, welches „kurz und auffällig genug ist, um sich einzubürgern" gefunden zu haben in Anspruch nimmt, in seinem einschlägigen Aufsatz *Symblysma oder der Überschwang der Jesuiten* sechs Autoren jesuitischer Provenienz zu, nämlich dem in Text selbst ungenannten Peter Browe, dann Peter Lippert, Hugo und Karl Rahner, Carlo Passaglia und Hans Urs von Balthasar, dessen Schriften für ihn wohl die meiste symblysmatische Relevanz hatten und der einen dem Symblysma entsprechenden Begriff geprägt hat; nämlich den der *circumincessio.*

Dieses Symblsyma, übersetzt das ‚gemeinsame Überschäumen‘, ist eine einprägsame Wortschöpfung, um einen zwar wohlbekannten, doch nicht wissenschaftlich anerkannten Prozess zu bezeichnen, nämlich das Wirken des Heiligen Geistes im Pfingstereignis. Rosenstock sieht seine Überlegungen über das Symblysma als einen Beitrag zur Biographik: es geht ihm um die bisher vernachlässigte „Geschichte ‚biographischer Gruppen'", welche als ein „Kapitel in der Geschichte des Heiligen Geistes" dargestellt werden kann. Dessen Wirken geht einher mit einer einschneidenden biographischen Wende im Leben der Betroffenen. Eine Gruppe, deren gemeinsame Biographie durch ein symblysmatisches Ereignis, also wechselseitige Begeisterung geprägt ist, erlangt eine besondere Art von Erkenntnis, eine Inspiration in Form eines verantworteten Wissens und einer verantwortlichen Wissenschaft.

Das Symblysma ist somit zugleich Rosenstocks Antwort auf den Bedeutungsverlust, welchen Pfingsten als gruppenbildende und sprachverleihende Inspiration im zeitgenössischen Christentum erlitt und welchen er wiederholt beklagt hat. So behauptet er in einer Rede: *Pfingsten hat schlechte Zeiten.* Angesichts dieser überragenden auch methodischen Bedeutung, die Rosenstock dem Pfingstereignis zuschreibt, ist es kein Zufall, daß im Kontext solcher Überlegungen der Name seines früheren Patmos-Gefährten, Karl Barth nämlich, und der Wunsch einer pfingstlichen Versöhnung auftauchen. Rosenstock schreibt: „Alle weltlichen Losungen des Parteigängertums müssen untergehen in der großen Wiederaufnahme (readmission) von Pfingsten; denn dies lehrt die wahre Vernunft." Ohne jeweilige und fortgesetzte Wieder-holung von Pfingsten schaffe nämlich, so Rosenstock in seinem Artikel *Pfingsten und Mission*, die Erinnerung an das ursprüngliche Pfingsten „nichts als Verzweiflung."

Das Symblysma bezeichnet nach Rosenstock die Polyphonie des Heiligen Geistes. Diese ist mit den üblichen biographischen Kategorien, die bei einem intellektuellen Aufeinandertreffen verschiedener Individuen angewandt werden, insbesondere mit dem von Biographen von Einzelnen gerne herangezogenen Gegensatzes von ‚recht haben' und entsprechend ‚unrecht haben', unvereinbar. Weniger theologisch und stärker soziologisch gesprochen: Rosenstock beschreibt mit dem Begriff Symblysma die im Wortsinne inspirierende, biographisch einschneidende Wirkung der dynamischen Auseinandersetzungen innerhalb von bestimmten Gruppen verschiedener Individuen und nimmt gleichzeitig das Christentum in seinen Aussagen ernst.

„Nicht der alleinstehende Mönch, nicht der einsam schaffende Genius, aber die ihrer Eigenart in der Vermassung treubleibende Einzelgruppe wird den Heiligen Geist bezeugen: vorübergehend und sterblich wie Ehe und Freundschaft, aber auch beseelt und durchgeistigt wie alle Pfingstgeschöpfe."

In einem solchen kreatürlichen Kreis geschieht Wahrheit, und zwar eine Wahrheit, die sich durch die Verantwortlichkeit des jeweiligen Individuums praktisch und öffentlich bewährt, einerseits gegenüber dem herausfordernden Ereignis, andererseits aber auch voreinander und vor allem auch vor einem äußeren Publikum. Manifest wird die Wahrheit im öffentlichen Sprechen, welches das Sprechen via Print in Büchern, Verlagen und Zeitschriften ausdrücklich einschließt. In *Out of Revolution* heißt es: "[…] inspiration, genius, talents, or simply thought, language, writing, are not merely means for starring lonely individuals; they are uniting people in a common life." Das Publizieren von Büchern wird somit zu einer Lebensnotwendigkeit, als ein Ergebnis eines überwältigenden symblysmatischen Ereignisses. In seinem bekenntnishaften Text *Ich bin ein unreiner Denker* heißt es:

„Es ist kein Luxus, ein Buch zu schreiben; es ist ein Mittel, weiterleben zu können. Indem ein Mensch ein Buch schreibt, befreit er seinen Geist von einem ihn überwältigenden Eindruck. Ein Buch ist echt, wenn es nicht willkürlich geschrieben wurde, sondern geschrieben werden *mußte*, um den Weg frei zu machen für ein weiteres Leben und Tun."

Das Symblysma als ergreifendes und produktives Ereignis steht bei Rosenstock für eine „Einheit des Geistes", welche auch mit Unzeitgenossen, d.h. „zwischen ganzen Epochen" möglich ist. Dem Anspruch ist es auch möglich zwischen verschiedenen religiösen Traditionen, ohne essentielle Gefährdung einer Seite, denn „das ursprüngliche Pfingsten [war] nicht ein Phänomen der Bekehrung, sondern eines der vielsprachigen Offenbarung […]." Symblysma ist Ausdruck einer sprachlichen Einheit, die mehr ist als die Summe der Sätze oder die Summe der Sprecher. In dieser Hinsicht ist das Symblysma eine Reformulierung von Rosenstocks früher Einsicht, daß die Sprache weiser ist als derjenige, der sie spricht.

Entscheidend an Rosenstocks Begriff des Symblysmas ist also die Etablierung eines gemeinsamen Geistes in einer

Gruppe von Individuen, der aber *wesentlich nach außen gerichtet ist und wirkt* (entsprechend wirkt das Pfingstereignis). Ein nach außen abschließender Geist ist Ungeist und nicht zukunftsfähig.

Als Unzeitgenosse, als Denker aus dem Anfang und der Mitte des vergangenen Jahrhunderts kann Rosenstock Zeitgenossen der zwanziger Jahre des einundzwanzigsten Jahrhunderts viel mitzuteilen haben – die folgenden Texte demonstrieren dies auf eine eindrückliche Weise. Nicht zuletzt zu nennen ist seine Lehre, daß es in Gesprächssituationen die Hoffnung auf ein Symblysma zur Gestaltung einer gemeinsamen wider-ständigen, aber friedlichen Zukunft gibt. Man muß diese Gesprächssituation nur zulassen wollen – es wären „time bettering days".

Knut Martin Stünkel
Ruhr-Universität
Bochum

INTRODUCTION

"In real life 2 + 2 never equals 4. Because we have to make sacrifices for each other not the slightest life process can even start unless the sentence 2 + 2 = 4 is thrown out the window."

EUGEN ROSENSTOCK-HUESSY WAS ONE among scores of German intellectuals of the highest caliber who in the 1930s turned their backs on Nazi Germany and settled in the United States, temporarily or permanently. With the help of a friend, Prof. Carl Friedrich, teaching at Harvard at the time, Rosenstock-Huessy in 1933 initially found a position there. In 1935 he relocated to Dartmouth College in New Hampshire where he remained as a professor of Social Philosophy until his retirement in 1957. Thereafter he lectured at various institutions in the U. S. and Germany. Throughout his life he was enormously productive as a scholar and writer.

In 1970 Clinton Gardner, under the imprint of Argo Books, a publishing entity in Vermont he founded, brought out two collections of essays or short papers by Rosenstock-Huessy, *I Am an Impure Thinker* and *Speech and Reality*. Gardner had been a student of Rosenstock-Huessy's at Dartmouth in the 1940s, and the emigre professor remained for him a lifelong inspiration. The two books together included twenty pieces by Rosenstock-Huessy, all copied from English language unpublished typescripts or, in a few cases, from previously printed originals.

Rosenstock-Huessy was alive at the time—he died in 1973— but he was failing and not engaged with this project of Gardner's, although Rosenstock-Huessy's companion, Freya von Moltke, certainly was. The seven pieces in *Speech and Reality*, the volume which appeared first, were grouped by Gardner on the basis of content, not date of composition, which stretched from 1935 to 1955. The thirteen pieces in *I Am an Impure Thinker* (which featured a foreword by W. H. Auden), written between 1938 and 1962, are not in any easily discernible order. Two of them are excerpts from Rosenstock-Huessy's major work in English, *Out of Revolution: Autobiography of Western Man* (1938).

The volume you are holding consists of English-language pieces also assembled by Gardner much later, in 1981, although with one exception, they too have never been formally published as separate entities. Nine titles in all, Gardiner's assemblage was preserved only in the form of a punch bound 8-1/2 x 11" compilation of photocopied sheets, about a half inch thick. The pages of the collection as a whole, ambitiously entitled "Rosenstock-Huessy Papers, Volume 1," were not even continuously numbered. The present book, then, may be thought of as papers that were left over after the publication by Argo of *Speech and Reality* and *I Am an Impure Thinker*, but the quality of the contents hardly fits the derogatory "left over." Some of Rosenstock-Huessy's most original and impressive work appears here. Parts of it, it must be said, are philosophically difficult and not easily comprehended. They present unfamiliar truths, requiring the reader to pause and re-read. Great originality has that effect. Rosenstock-Huessy was never gratuitously obscure as a cover-up for nonsense. He strove to be understood, and his thought is always well grounded, but his thinking constantly leaped ahead..

It would be nice in an introduction like this to place each of these pieces in a fuller context of Rosenstock-Huessy's *oeuvre* as a whole: what was first written in German and re-worked by him as a text in English; what was first drafted in English and re-written for a volume in German; locating where in his complete corpus one finds the seed of an essay; noting what appeared earlier and was substantially modified, and so forth. We have, however, done none of this bibliographical research. The reader will find much of it done for us by Lise van der Molen in his indispensable *Guide to the Works of Eugen Rosenstock-Huessy: A Chronological Bibliography*...(Essex, Vermont: Argo Books, 1997; available online at: erh.fund.org), but mostly such an intellectual bibliography, a treatise on the development of Rosenstock-Huessy's thought, remains a task for future scholars.

More than forty years after this collection was conceived, and as it happens fifty years after Rosenstock-Huessy's death, it is surely time to properly present this treasure to the world. Following the original, the nine works included here are presented simply in order of composition 1937 to 1961. It is immediately apparent that the collection is dominated by Rosenstock-Huessy's intense interest in the proper role of natural science in the post-Enlightenment western world. We are all so grateful for the astonishing achievements of physicists, biologists, chemists, engineers of all sorts, etc., in reducing the burden of primitive existence and extending knowledge, that science has taken on an outsized or out of balance role in human estimation. Men (and women) are normally polytheists in Rosenstock-Huessy's assessment, and among the many modern gods that we worship, none may be more unquestioned in this pantheon than the idol of science.

The problem with science or technocracy is not that it is not a good in itself, but that it is only one good among many and that it always has a terrible underside: nuclear energy, nuclear weapons; genetic cures, genetic modification; social media, social manipulation; agricultural bounty, devastating pesticides and polluted earth and waters. The list is long. In Rosenstock-Huessy's four-sided grammar of existence, science is categorized as part of the realm of the objective outward approach to space and matter. Equally important when looked at fundamentally is the inner space, the subjective or the subjunctive, and in the realm of time, the imperative, which moves us into a better future, and the "trajective", which guards and revives the necessary past. A scientific statement may be valid and true, but it may also be limited, only a small part of a larger truth. In my personal estimation, no part of Rosenstock-Huessy's life work has greater implications for the future of our wellbeing than his thinking on the proper role of "science" in the broadest sense, including technology. This claim may sound far-fetched, a call to return to imprecision, fuzzy speculation, guess-work, ignorance rather than "truth." Reader, study this book and you will be convinced. Science is presented here as a passion in contrast to the dryness of its formal presentation. Both the psychology and the sociology of the scientific endeavor since the 16th century is revealed. None of this takes away from the merit of the bourgeoning of science in the West since the time of Galileo or, in this book, since the time of the "founder" Paracelsus, who has a starring role along with Faraday, the much later "classic" researcher.

The rise of "science" is paired by Rosenstock-Huessy with the invention, or discovery, of "nature," a word, one may think , that has no history. In Rosenstock-Huessy's view, nature is that part of the universe that cannot speak and for the most part cannot listen. It is telling that the nature idolatry of the Enlightenment has failed to protect the creation against horrific abuse by humankind, although environmental disaster is not a theme in this book. The book does tell us, in part, however, how we got to this point. Rosenstock-Huessy's deconstruction in chapter five of the seemingly bland and innocuous *title* of Sir Arthur Eddington's *The Nature of the Physical World* (1928) is a startling revelation.

Although restless and a maverick in the academy, Rosenstock-Huessy was a university teacher most of his life. He wrote extensively about the meaning and process of education at all levels because he was immersed in it, including, early in his career, so-called adult education. Education, its very meaning, has a prominent place in this book, and it is hard to think of any writer who was more original and profound on this topic, employing none of the usual conventional themes. John Dewey, perhaps the most influential theorist of education in 20th-century United States, was for Rosenstock-Huessy a particular *bête noir*. He loved poking sacred cows like Dewey, quite fearlessly.

Time is the cradle of education, with the student of any age learning for the future and the teacher imparting the known from the past. Rosenstock-Huessy's preoccupation with the meaning and nature of time in multiple contexts is another basic subject introduced here with startling freshness. Nothing concerned him more than the problem of generational conflict and the essential need always to reconcile fathers and sons, to use the common (or biblical) metaphor, or how to turn "distemporaries" (Rosenstock-Huessy's coinage) into contemporaries. We are talking now not about family squabbles but about world peace, the ultimate aim for people everywhere. The analysis of St. Augustine's *De Magistro*, in the third essay, "Man Must Teach," links time and a proper understanding of education. Augustine's illegitimate son, Adeodatus, who needed to be educated, figures prominently in it.

We have named this collection after the title of one of the essays, "Time Bettering Days," a phrase from Shakespeare, because that essay so well exemplifies Rosenstock-Huessy's range and originality as a scholar always attuned to social improvement. The essay is, among other things, a remarkable display of learning—Rosenstock-Huessy was not above the vanity of scholarship, intimidating the poor reader—but its message relates also to "science," i. e., mechanized uniform, undifferentiated time—tick, tick, tick, tick, tick—versus time creatively broken up by human beings into moments, hours, minutes, years, holy-days, commemorations, celebrations, mourning, protocols, time redeemed by human choice, rescued from time defined as the fourth dimension of space, an absurd diminution of our most basic milieu. For human beings, space is mere background; time is the stream in which we live, with past, present, and future vital human inventions and not the least bit mechanical. The present is not a razor's edge; it is what we make it: the present moment, the present year, the present century, the present millennium, and Rosenstock-Huessy definitely thought in the context of millennia, with a full grasp. He published a book with the title, *The Christian Future: The Modern Mind Outrun*, and another with the subtitle, *Autobiography of Western Man*.

He was a committed Christian and a brilliant and unconventional interpreter of Christianity. Few historians have seen so deeply into the influence of Christianity in Western and world history, often not easily evident. In the *Christian Future* he spoke of "Christianity incognito" where the inspiration of Christianity has to be revealed, and the actors are hardly aware of their heritage, i. e., of the spirit under which they are striving. He says here boldly in one essay, "all true Christian religion is secular just as its founder." Another essay herein celebrates a new translation of Calvin's *Institutes*, by Ford Lewis Battles, a friend of Rosenstock-Huessy's. In it, the problems of predestination and eternal damnation are tackled head on by Rosenstock-Huessy's method of always searching for deeper meanings and asking what is really at stake here.

Be prepared, then, for a feast of rare insights, originality on every page, bold assertions, passionate discourse with the claim that our very survival is at stake. Heed one of Rosenstock-Huessy's favorite injunctions: Listen, Reader, lest we die (*Audi, lector, ne moriamur*).

A NOTE ON EDITORIAL METHOD

OUR PRINCIPAL GOAL IS to preserve and make available to all who may be interested a small selection from the extensive writings of Eugen Rosenstock-Huessy. The papers or essays that follow have been passed on to the editor without much context concerning them. Each has an individual history, which it would be difficult to trace. If they were originally in longhand, we don't know who then typed them. We don't know how much the typist "edited" or "corrected" Rosenstock-Huessy's original manuscript. We don't know to what degree Rosenstock-Huessy himself reviewed and revised various typescripts, although there are occasional script insertions in his hand. If there is more than one version, we don't know which of these Rosenstock-Huessy considered the "final" approved copy.

That is one set of problems presented to any editor intent on publishing these papers. Another is that although Rosenstock-Huessy was a brilliant, imaginative writer, and although he learned to speak English at an early age, English was not his native tongue. Hence, the prose is sometimes awkward or unconventional in style, although there are very few undecipherable passages with regard to meaning. In some respects he was a dazzling stylist. In any case, we have not tried to rewrite this text to make it clearer where it seems opaque. Preservation is more important than instant clarity in a collection like this from an author who died a half-century ago. Be assured, Rosenstock-Huessy never spouted nonsense, but he is often deep, complex, and surprising. Where we can be helpful to reader understanding, we have added a footnote or inserted an explanatory word in square brackets.

In the matter of punctuation, there is little standardization throughout. Commas are strewn on the pages as though placed more for guidance in an oral presentation than to help a confused reader. We have not attempted to alter this errant punctuation, except to remedy obvious typos and to standardize some practices. We use four dots for ellipses at the end of an incomplete sentence, and three dots in the midst of a sentence, with the proper spacing. If an errant comma obfuscates a sentence, we may silently remove it, but rarely; we also very occasionally add a comma for clarity. We also follow British rather than American practice in placing punctuation, a comma or a period, outside of quotation marks, which is the British practice that Rosenstock-Huessy *mostly* followed. The standard goal of copy editing, consistency, is not our aim, and we have ignored it as a rule. The point of the foregoing is that a copy-editor working with a living author not as important as Rosenstock-Huessy would have a merry time "improving" the text line-by-line and hoping for the author's applause and consent. But Rosenstock-Huessy is a highly important author whose exact words may be described as precious, and he is not here to accept or reject changes to his prose. Hence, in the spirit of professional *documentary* editing our procedure has been to change almost nothing of what has come down to us, and equally important, to signal to the reader virtually every change we have made. For example, we have usually bracketed the insertion of any word or punctuation point, or if not that, spelled out any alteration in an editor's footnote. The only silent alterations we have made in the text are the correction of obvious typos, which are not infrequent—e. g., "wordly" intending "worldly," or "test" intending "text," or "sunce" intending "since". Sometimes, too, we have silently corrected typos in proper nouns. Where Rosenstock-Huessy quoted from a well-known source, such as Shakespeare, and the quotation is not perfectly faithful to the original, we have changed the text to conform to the standard source. If there is any uncertainty in these matters, the change is footnoted, preceded by "*Ed.*"

Rosenstock-Huessy often used British spellings, which we have not altered—coloured, enamoured, centre, etc.—and in general, as noted, we have ignored the usual copy-editing rule of bringing about any consistency in spelling and capitalization, which may be quite irregular within a single essay or within the book as a whole. Rosenstock-Huessy sometimes used capitalization only for emphasis, not as part of a rule. He sometimes used archaic or obsolete spellings— "summersault," "retractation" for "retraction," for example—which we have left as is. We have retained British spellings, but not the practice in Britain of using single quote marks initially. Here we have followed the American rule of double quotes initially, and for a quote inside a quotation, single marks.

The typewriter, as contrasted to the computer as a writing tool, or for that matter as contrasted with a linotype or monotype machine, was a rigid instrument. To signal italicization the typist's only option was underlining a word. We have silently converted underlined words and phrases to italics. We have also italicized book titles, in accordance with standard practice, and sometimes words and phrases in foreign languages, and we have converted hyphens into em dashes where that is the author's obvious intention. It is impossible to make a true em-dash on a typewriter.

Rosenstock-Huessy's numbered footnotes, which with the exception of those in the first essay in the book are very few, have been edited to bring them into line roughly with the *Chicago Manual of Style* and with standard citation style in general. Abbreviated titles and author's names have been expanded, and place and date of publication added where called for. On the other hand, we have not expended hours of time trying to make more precise a generalized citation, e. g., "Nietzsche, *Werke*, XVI, 359". We have also not verified page references in the works cited.

Paragraphing in the original typescript is often erratic or arbitrary, but we have not, for the most part, tampered with it. In a few cases some adjustment is called for, and we have done it silently, in particular adding a paragraph break where it is obviously needed. Our general rule is: If there is sufficient clarity for the reader's comprehension, we have not meddled.

The above relates to editing *within* any one of the nine pieces herein. We certainly have not attempted to bring about consistency in the book as a whole. Some of the essays herewith are rougher than others. Some have many obscure passages; others are perspicuous throughout.

Finally, the editor wishes to recognize here the valuable contribution of Lynn K. Jones to the preparation of this work.

Norman Fiering

One

A CLASSIC AND A FOUNDER

Part I. The Classic of Science. The Scientific Grammar of Michael Faraday's Diaries

THE GRAMMAR OF HIS DIARY

For sixty years, the managers of the Royal Institution of Great Britain held in their care the daily manuscript records of the researches of Michael Faraday (1791–1867), leading physicist and chemist of the first half of the nineteenth century. The sheets cover more than four decades.He was in the habit of describing each experiment and every observation inside and outside his laboratory, in full and accurate detail, on the very day they were made. Many of the entries discuss the consequences which he drew from what he observed. In other cases they outline the proposed course of research for the future. Thus this diary is supplementing our general conception of science. We sometimes are inclined to look into a science not our own as into a catalogue of results. In Faraday's Diary, it becomes again what it really is, a campaign of mankind, balancing in any given moment, past experience, present speculation, and future experimentation, in a unique concoction of scepticism, faith, doubt, and expectation.

Therefore, our interest in this diary lies quite outside the range of propositions and proofs for any specific content or aim. It centers round the logic of Faraday's mind, round the method of his strategy, both in thought and experiment. Seven thick volumes were printed a decade ago handing over to the general reader this diary for general use. What might seem merely a physicist's special theme, really may be used as a symbol of the true passions of the human mind. An experimental logic derived from this and similar documents will show that all Greek logic is an abstraction void of the sense for time. We never reason in the void of timelessness. Faraday thinks from day to day, against a background of older thinking, and anticipating new facts of tomorrow. In other words, he thinks in three dimensions of time: past, present, and future. Scientific logic becomes meaningless, when we dissect it and analyze any one of its statements or conclusions outside the interplay of past knowledge, future experimentation, present day speculation. The famous doubt of the scientist is the shadow cast on the past by the expectation of future better knowledge. Without this relation it would be sterile. Experiments are based on what he knows already. Finally he speculates because he has to pause between future experiment and previous knowledge. To reestablish the elementary fact that the human mind cannot think except in the three dimensions of time, is one of the most burning scientific needs of our age, so that the centuries of pure physics may be continued by an equally successful series of biological centuries. As long as we talk of the dimensions of space only, and use the obsolete and wholly unrealistic Greek and Roman tools of logic, the biology of the human mind remains under the spell of an irrational conception. Faraday, then, by his untiring faithfulness in keeping his diary, contributes to our understanding the objects of his scientific research in magnetism, electricity and light, but he also makes us understand the scientist himself, as a living subject, the mind in action.

The questions which we had in mind when we analysed the seven volumes were, for example, what was the driving urge behind all the steps of all the experiments.[1] The wonderful humility of Faraday—he never sought distinction in society, always kept the faith of his Sandemian friends, a good Christian sect of poor people[2]—makes it possible to discard all external causes or motives. His inner desire, then, what was it behind so many failures and so few successes? For in forty

1. *Ed.* Prof. Rosenstock-Huessy credited seven students in his course on "University Life, Past and Present: A Philosophy of the Sciences" with helping him analyze the diaries, in particular "Mr. Symmons from Phoenix, Arizona, [who] did most of the work."

2. *Ed.* The Sandemanians were a Christian sect that originated in Scotland early in the eighteenth century.

years the blunders, mistakes, miscalculations, and wrong hypotheses far outnumber the lucky shots. This fact is so impressive, that it leads to a more general recognition: Sciences when not treated as a catalogue of results but as a process of collective action, are in fact a systematic and voluntary relapse into errors. "We must allow the scientists to err", said Pope Leo XIII. Science as a process is the organization of all thinkable errors in order that, as a later result, error may be overcome. No shepherd could survive, if he made one hundredth part of the blunders Faraday made during his life. A shepherd's life can hardly forbear more than five percent of error and experiment. It takes the complete isolation of a laboratory to give us the privilege of making mistakes at random. Now this voluntary creation of a maze would be unexplicable were it not for the anticipation of something behind the confusion which is apt to reward us for this voluntary relapse into ignorance. The scientist is like a man who purposely marches many steps backward before he jumps a trench. Scientists— in the midst of their experimental avatar— must be able to know less than common sense and everyday technique take for granted already. Why? Because they anticipate some unknown element outside our present day knowledge that will prove the narrowness of its diameter. Against a too narrow circle, their vision tries to enlarge the facts of interplay, relation, dependence and interaction. More unity of nature we may call the dogma of science; in this formula, unity is nothing absolute, and it has to be compared to the previous opinions on unity before it makes sense. The logic of science is a relative logic of infinite approximation. It increases relations, it unifies twenty experiences by chaining them to the triumphant chariot of systematic experimentation. Fifty guinea pigs investigated one after the other, cease to be fifty cases of murder. Fifty acts become one unified effort; in this way, experimentation is absolutely different from mere experience: It organizes experience by anticipating unity.

As Faraday exclaims: "Surely this force (gravity) must be capable of an experimental relation to electricity, magnetism, and other forces, so as to bind it up with them in reciprocal action and equivalent."

In this realm, then, of creating unity, Faraday speaks, as any complete human being, all the three languages of emotion, command and narrative. The emotions are those of wonder, admiration, and doubt or, against doubt, emphatic assertion. The imperatives are directed to himself. The narratives fix experiences.

Faraday's scientific grammar with regard to the imperative is simple: "I must look at Weber's result to see how they build in with these considerations and what the results are." Later he says: "Astonishing how great the precautions that are needed in these delicate experiments. Patience. Patience." Probably a rare entrance in any man's diary, because so few people allow it to contain more than descriptions or analyses of feelings. Again he writes: "Want to try a mass of something to ascertain whether it will sensibly affect the directions of the lines of force of the earth—that it may approach a step to the action of oxygen." In reference to an experiment already undertaken: "Have arranged a check—shall make this adjustable by hand. It is an important adjunct in experiments of observation." As an aid to his poor memory he frequently says: "Query these results." Or: "Remember the dip." and: "Must clear all this up by further experiments." He may write: "The hypothesis is not so much mine as one renewed from old times. Look at Euler's letters and what he says. Look for cases to prove it." These Imperatives directed to Michael Faraday only lead up to more general rules of wisdom: "Let the imagination go, guiding it by judgement and principle, but holding it in and directing it by experiment." And the grammatical form of the imperative is not even used in this comforting sentence: "To point out or lead to a knowledge of what it either cannot explain or has not explained, is quite as important for the progress of knowledge as to establish what it can do." In the quotation on the unity of gravity, electricity and magnetism quoted above, he ends with a remark that is equally general and personal: "Consider for a moment how to set about touching this matter by facts and trial."

Since the diaries were kept primarily for Faraday's own benefit, they frequently betray his emotions of wonder and surprise. Thus: "I have been analyzing certain experiments in reference to the notion that gravity itself may be practically and directly related by experiment to the other powers of nature and this morning proceeded to make them. It was almost with a feeling of awe that I went to work, for if the hope should prove well founded, how great and mighty and sublime in its hitherto unchangeable character is the force I am trying to deal with, and how large may be the new domain of knowledge that may be opened up to the mind of man." Later he says: "After all, there is much which renders these expectations or similar ones hopeless: for surely, if founded, there must have been some manifestation of such a condition of the power in nature. On the other hand, what wonderful and manifest conditions of natural power have escaped observation, which have been made known to us in these days." When something unexpected would come of an experiment, his excitement would be intense: "But now came forth a new and striking result. Strange! Must find out the cause of this. What effect does this force have in the earth? His experiments meant more than technical proof to him: "It is exceedingly beautiful to see in all these arrangements how beautifully the lines of force represent the disposition of magnetic power." Or "Such beautiful delicate indicating curvatures." "The results are beautifully near and proportionate." Words of emotional description frequently used were "astonishing, I durst not, excellent, it was not easy because of imperfect eyesight, interesting, remarkable, curious, I begin to despair."

In his scientific grammar, certainty and doubt, naturally alternate. "Surely this force must be capable . . . ," he said in the sentence on gravity. Of some conclusion reached he might write: "Hence this method seems defective in principle, or at all events in sensitiveness; and yet it is very sensitive. Certainly there was no hopes for any optical results since there are none here. I think Plucker must have been mistaken in his result and that my old observation was right."[3] And again: "I think that I may trust the reality of these negative results." At times he is quite positive and says: "I have no doubt," or "I have proof," or at least, "from all these experiments, I am led to conclude." More often than direct questioning he replaces his own conclusions, obviously vivid in his own imagination, stolidly with an appeal to the judgement of others. "I refrain from extending these views, as might easily be done, to the atomic theory, being rather desirous that they should first receive the sanction or correction of scientific men." Or "I have refrained from all reasoning on the probability of the compound nature of nitrogen or upon what might be imagined to be its elements, not seeing sufficient reason to justify more than private opinion upon that matter."

In the laboratory, the description of his experiments is more or less a sequence to his own arrangements. Evidently these descriptions

3. *Ed.* Julius Plücker (1801–1868), German mathematician.

then, are no pure, unpremeditated narratives. All laboratory-facts are man-made, i.e. secondary experiences. Lest we exclude the best and most immediate source from which to know his power of narrative, and the delicate way by which a vigorous impression was transformed into expression, we must turn to the pages of the diary where he tells of unexpected phenomena in the street, or on the sea shore. Of course, they are much longer than the short imperatives; however, we should keep in mind that, in the system of thought, one short command, "patience", equals a long tale about the past. Retrospection is bound to be long; the plunge into the future is its very opposite. Here, then, follow some examples of descriptions: Experimenting one day with chlorides he writes: "Not with Magnesia; only chloride and proto-chloride produced. There was a fire on Thursday evening in Broad Court, Anny Lane. The clouds were low and received a strong illumination from the fire beneath them. The angle taken from the top of the Royal Institution by a quadrant formed by the clouds, the Institution, and the fire was 24°. Hence the height of the clouds will be . . . equal to . . . ." Again spending the day in the laboratory deep in chemical analysis he says: "Phoenician coin analyzed—is composed of copper and silver. It was a small cast coin weighing about 120 grains, having a rough white surface but brittle coppery fracture. It contained no lead, tin or antimony. The design was bold and well preserved and consisted apparently of characters or symbolic marks. A whole bag of these coins were found at [?] and were bought for a pound." Still another day he walks out of his laboratory and sees: "At Folkestone the atmosphere clear and fine view of the cliffs of Dover. Soon after sunset (the wind being about S. S. W. so as to blow on land) observed a cloud forming just the brow of Shakespeare cliff. It streamed inwards, increasing in size, but all seemed to pour nearly from the same spot; the air which came from over the sea there taking on a visible form and passing in to the interior as a cloud. By degrees the generation of clouds took place along the whole line of cliff from Dover to Folkestone hill, the wind still carrying the portion formed over the land. We ascended the cliffs about ½ a mile beyond Folkestone hill about an hour after sunset and found all above developed in dense, moist mist, so as to deposit water on our clothes; the temperature also low to the feelings. We walked back towards Folkestone and on descending a little way down the hill by the road emerged from the cloud and found all clear beneath. The cloud was extended a considerable way in land, covering the tops of the hills. Was not this effect produced by the cooling of the surface of these hills after sunset by radiation into the clear space above, and the consequent cooling of the moist air brought by the wind from the sea below its point of deposition?" Again the next day, his lack of departmentalization allows him this entry: "At times when the wind has been rather strong, I have frequently watched the gulls who were flying over the waves looking for food, and have often seen them move slowly against the wind or remain stationery facing it, balancing themselves on their wings but without flapping them. This has lasted for 1, 2, 3 or more minutes, and I think could not be due to any previously acquired momentum because they would suddenly sweep round, going down with the wind, and then again return against it, all without flapping the wings; I have also remarked hawkes over land advance in a similar manner in similar circumstances, without having been able to detect any motion of the wing calculated to support them. They seem to remain suspended in the air by an apparent balancing of the body on the wings against the wind. How do these birds fly? And why may not a man or a machine fly in the same way in the same circumstances?" A year later he returns to the same place. And again has the opportunities of remarking the balance of the gulls in strong wind: "Many of them would rise together and there seemed to be a sort of emulation among them; all had there heads to the wind which was here parallel to the cliffs. . . . Perhaps the effect which may sometimes be observed in flying a kite may be connected with this subject. Sometimes a kite when badly rigged will, upon rising, not cease to ascend when the string forms a certain angle with the current of air, but will continue to mount, taking nearly a horizontal position in the air, and that till the string is nearly vertical when the kite generally falls over and comes down."

At yet another time while, in his laboratory making experiments on light "and then oxide of zinc seemed fixed and unchanged by the high temperature produced." Suddenly this paragraph is inserted: "John and George Bonnard being in a hay field where many large cocks of hay were, had occasion to notice the effect and progress of a powerful whirlwind; it took up the whole of a hay cock, raising it in the air, whirling it around and expanding it over a space 6 or 7 times its original diameter and then letting it sink a little in advance on the neighboring ground or trees. It is evident that the progressive motion of this whirlwind (and the same with most of them) was not due to the advancement by a general wind of that portion of air which was first put into rotation but that of a general mass of air; nearly quiescent, contiguous portions assumed the rotating motion in succession, so that when the air over a haycock had rotated and taken up the light matters beneath, its motion gradually ceased whilst the neighboring parts revolved and the just raised hay fell again." Again later, he leaves his laboratory: "This evening a magnificent aurora borealis occurred. At 11 o'clock it was like a powerful clear twilight or the break of morning from behind a low ridge of dark, picturesque clouds towards the North West to East North East and 40 or 50 degrees in height. Sky otherwise clear, wind from the south west but slight in power. . . . A fine, broad pillar of red light gradually formed . . . after innumerable changes the light both as to color and intensity, the whole gradually assumed the appearance of faint columns or rays . . . dancing or flashing perceived. It appeared as if part of the sky towards the zenith suddenly glowed with a phosphorescent light. . . . A remarkable fact relative to the lines of direction toward the one spot south of the zenith was that, even when the blush did not proceed along them, but across them or simultaneously over a large space, still they were visible and apparently as fixed in their position as ever." And once more: "A beautiful aerial phenomenon observed about St. Paul's Church, from the shadow of the dome, and the part above cast on very thin clouds moving at that height. The moon at full and rising. . . . The effect was very beautiful. Many persons went away fully convinced that rays of darkness were issuing from the church. Time about 8 o'clock."

The classical case for this respiratory process between experience and experiment occurred when a friend gave Faraday a large Leyden jar. It was broken by a shock of electricity in an experiment. Instead of bewailing the loss and discarding the jar, he proceeded on an intricate series of new experiments to determine why and how the electricity broke the jar. He made drawings of the break, and though thoroughly excited by the accident, he conducted his series of investigations as if he never had planned anything else.

William Blake called division the sin of man; Faraday was a great man because he was utterly undivided. His whole, very harmonious, very well balanced, to be sure, still his whole nature, and not a brain, a slave of the intellect, was at work through the years; though we owe his diary, partly at least, to his one weakness, his unreliable memory, it reflects the rare character who immerged completely, soul as well as body, into the intercourse with his world, and used the brain in the limited way in which it is useful, and for those ends for which it is given us. On the basis of imperatives, emotions, and narrations, he built up his few but precious speculations. Their simplicity rivals with their forcefulness.

Words frequently used to express doubt and speculation were: "it has occurred to me, perhaps I am in error, it would appear, upon consideration, I suspect, would this imply, I think, I believe, a correction needful, at times it seemed so, it is not sure, I want clearly to understand, suppose that this were so."

"I am learning how to observe." "I have not found it so." "The point will require investigating." "This does not accord with the facts; but I want more and more distinct results, and only reason thus to preserve under the disadvantage of a sadly failing memory the ideas that I may want to reconsider hereafter. The facts, as far as they go, are I believe good." He balances his explanations even in the moment he is formulating them for the first time: "Many interesting points would arise here for consideration. . . . Is the diminution permanent or is the full charge restored on lowering the temperature? Either answer would be important in the consideration of the nature of steel magnetic charge." Or he faces the negative: "I think that I may trust the reality of these negative results. The whole day almost in vain; for after the end of it all discovered a source of error which vitiated all the results and also those of yesterday—but it was well to know the error. No wonder the results of yesterday were incomprehensible." Or this: "So now I believe that all the effects I had heretofore obtained were due to the falling or rising loop of wire and not to any effect of gravity. At all events, we are purifying the inquiry from interfering causes." Cancelling his efforts, he might write: "Of a sudden all wrong and I see not why."

We gave his statement on gravity before: "Surely this force must be capable of an experimental relation to electricity, magnetism, and the other forces, as to bind it up with them in reciprocal action and equivalent effect." This faith in the unity of the elements composing different phenomena is called today, with an understatement, working hypothesis. The term is not exact; because it suppresses a number of essentials that such a faith must contain in order to make people work. It is, then, not a hypothesis for the objects, but an imperative for the subjects who do research: *It makes them work*. Again, it does not make work one man or another; to the contrary, such a subjective assumption is not the faith required by science. It must be a faith that may be shared by many, eventually by all scientists. For that purpose it must be in accordance with the main dogma of science: unity of nature behind all the phenomena. And even here the faith does not end. It must reach people not as individuals, one, ten, a thousand. It must make them cooperate in an integrated division of labor. We use the word faith rightly in all instances where people of different thinking and convictions cooperate. A child and his father, a police man, a farmer and a scholar, may have the same faith, though this faith is reflected in their brains in completely different concepts and words. Science is able to make cooperate catholics and mechanics, students and Nobel prize winners, because a common faith distributes the functions of workmanship despite all differences of rational formulation.

Faraday was a classic because the faith into the unity of nature came to him not as a heresy, but as the precious acquisition of two centuries, with the certainty of a social code, embodied by his master, discoverer and promotor, a member of the best society, Sir Humphrey Davy. The son of the blacksmith who was Michael Faraday, was not asked to fight the prejudices of the upper classes; he was invited to share and to advance their living faith and their most sincere and valuable endeavors. This fortunate constellation produces the classic, the type of man who is allowed to add to the trends of his times the integrity, strength and harmony of one especially well organized individual. We shall see, in the second case here under consideration, how unique Faraday's position was, how rarely society and individual are in the balance embodied by Sir Humphrey's pupil.

It needs scarcely saying that in our own days, scientists begin to assume so much power that they are threatened by the same cancer that kills any powerful group or clergy, simply by imparting power.

The classic serenity of Faraday is equally far distant from dawn and sunset of the day of science. By the absence of any fighting element in his mind, of any attack against the pre-scientific age, or of any self-defense of professional claims for power, in the whole diary,

Faraday's life proclaims the hours before noon when the domination of the new sun is ascertained already; however, the zenith of science is not quite completely reached, the light is still united and concentrated, not diffused in the thousand colors of the afternoon sun.

Here are some more short expressions of Faraday's faith. "No doubt a larger law of action would bring both or all three cases under one expression, but still that would not as yet show that bismuth is diamagnetic." Or: "Still, I think there must be some relation between these functions of light and electric forces." Again, he speculates: "Universe magnetism. Earth, Sun, Moon, probably all lie as mutually related magnets in common medium of space. In view of media, may very well speak of atmospheric magnetism in relation to earth."

"This space or state of space is new to our knowledge. So also is the space filled with lines of force new to our knowledge, i. e. to the knowledge of philosophers generally." About another phenomenon he muses: "Time in relation to magnetic force—probable existence of a medium; if time concerned, it will most probably be exceedingly short like that of its relation to light, and so perhaps for ever remain insensible to us." "If considering the reasons before given, there be the least hopes of finding the time, these hopes ought to be verified or exhausted. Can that be done thus?"

And so we are led on to two utterances; one is connecting the whole universe of man's mind—and let it be clear that the problem now is enlarged from the different departments in the individual mind of Faraday, emotions, dreams, volitions, memories, and ideas, to the more complex stage where mankind must survive as undivided whole, with science, art, religion, and legislation as immense units and organs of life. And the other is bringing together the external universe into one dynamic system, united in the way Laotse spoke of the unity of the wheel produced by the one point in which there is no wheel.

1. "If there should be any truth in these vague expectations of the relations of gravitating force, then it seems hardly possible but that there must be some extraordinary results to come out in relation to celestial mechanics—as between the earth and the moon, or the sun and the planets, or in the great space between gravitating bodies. Then, indeed, Milton's expression of the

sun's magnetic ray would have a real meaning in addition to its poetical one."

2.   "The Aurora borealis may now become connected with magnetic disturbances and storms in a very distinct manner; and if the variations of the atmosphere cause both, it will also tie both together by a common hub."

The last paragraph of Faraday's daily report on his work bears the figure 16,041.[4] And one of his last public utterances was: "for all the phenomena of nature lead us to believe that the great and governing law is one."[5] 16,041 and One—this is the great paradox of his life, faith and grammar. "When we consider the life work of Faraday it is clear that his researches were guided and inspired by the strong belief that the various forces of nature were inter-related and dependent on one another. It is not too much to say that this philosophic conviction gave the impulse and driving power in most of his researches and is the key to the extraordinary success in adding to knowledge."[6] As to 16041: "A good experiment would make him almost dance with delight." And as to One: "The Contemplation of Nature and his own relation to her, produced in Faraday a Kind of exaltation."[7]

THE THREE DIMENSIONS OF TIME

It will be our final task to establish the "respiratory process" between the 16,041 and the One as the most important contribution of the diaries to our understanding of the mind in action. For 16,041 reasonable doubts, we may say, were experienced, considered, tested and cleared against the background of One faith.

Before deepening his meaning of his respiratory process, we must listen once more to Faraday himself. For he knew that the mind

4. *Ed.* Faraday's Diary is a bound and sequentially numbered set of books, containing 16,041 numbered entries, from August 1832 to March 1860.

5. Edward Livingston Youmans, *et al.*, *The Correlation and Conservation of Forces: A Series of Expositions* (New York, 1865), 376. See further, William H. Bragg, *Faraday Centenary Exhibition* (London, 1931), 22, 25. J. H. Gladstone, *Michael Faraday* (London, 1873), 123ff "His method of working."

6. Royal Institution of Great Britain, *Report on the Faraday Celebrations, 1931* (London, 1932), 39.

7. John Tyndall, *Faraday as a Discoverer* (London, 1870), 186.

in action, his own mind, differed from the mind outside the body of science. "What a weak, credulous-incredulous, unbelieving-superstitious, bold-frightened, what a ridiculous world ours is, as far as concerns the mind of man. How full of inconsistencies, contradictions and absurdities it is. I declare that taking the average of many minds that have recently come before me (and apart from that spirit which God has placed In each) and accepting for a moment that average as a standard, I should far prefer the obedience, affections and instinct of a dog before it.[8]

Therefore we should try to view his lucid and keen mind against the society in which he as a scientist had to live. In his later years, a committee inquiring into the state of education, asked him, with many distinguished scholars, to express his opinions on the best training of the mind. The report, long forgotten, would deserve a complete reprint. Since our specific purpose is to show the isolated existence of a "classic", in the midst of the society of his day, one paragraph may suffice.[9]

Faraday stated that he had not the "training of the mind" usually expected from regular education in the classics and continued:

"The phrase 'training of the mind' has a very indefinite meaning. I would like a profound scholar to indicate to me what he means by 'training of the mind' in a literary sense, including mathematics. What is their effect on the mind? What is the kind of result that is called 'the training of the mind'? Or what does the mind learn by that training. It learns things, I have no doubt. By the very act of study, it learns to be attentive, to be persevering, to be logical, according to the word 'logic'. But does it learn that training of the mind which enables a man to give a reason, in natural things, for an effect which happens from certain causes: or why, in any emergency or event, he does, or should do, this, that, or the other? It does not suggest the least thing in these matters. It is the highly educated man that we find coming to us, again and again, and asking the most simple questions in chemistry and mechanics; and when we speak of such things as the conservation of force, the permanency of matter, and the unchangeability of the laws

8. *The Letters of Faraday and Schoenbein*, ed. by G. W. A. Kahlbaum and F. V. Derbishire (London, 1899), July 25, 1853.

9. Edward Livingstone Youmans, ed. *The Culture Demanded by Modern Life, a Series of Addresses and Arguments on the Claims of Modern Education* (New York, 1869) 463.

of nature, they are far from comprehending them, though they have relation to us in every action of our lives. Many of these instructed persons are as far from having the power of judging of these things as if their minds had never been trained."

Finally, in his observation on Mental Education, Faraday himself turned toward, the analysis of scientific judgment. He showed the beauty of "errors" if they were to be considered honest efforts between a dark, ignorant past and a more enlightened future, and defined error as "a presumptuous judgment", rendered too early.[10]

We are now, I think, in a position, to state our most important result.

THE SCIENTIST'S FISSION IN MIND AND BODY

In the grammar of this scientist, doubt, reasonable and experimenting doubt, retains its place between the great certainty with which he marches into the future and the seam free aloofness towards the past and its social routine. The scientist is freed from the responsibilities for routine and repetitive work. In the case of Faraday, this delegation of an experimenting mind by society worked beautifully, because his loyalties towards this same routine—society, his certainty of faith into a promised future, and his equanimity in his present stage of doubt, all were in perfect balance. Our faith into the future plus our loyalty towards the past are the parents of legitimate scientific doubt. This parenthood separates organized, scientific doubt from all scepticism or cynicism. It reveals what any "present time" of a civilization or a man really is. The present time is not the result of the past nor is it the "cause" of the future though this is the most current fallacy of our era.

This deserves our special attention. In natural science, it is true, the objects are treated as if the future depended on the presence, according to the famous formula of Laplace: "We ought then to regard the present state of the universe as the effect of the anterior state and the causation of the one which is to follow."[11] Only, what is true for

10. *Lectures on Education before Prince Albert delivered at the Royal Institution of Great Britain* (London, 1854) especially p. 47.

11. Pierre Laplace, *Theorie Analytique des Probabilités*, English translation 1902, p. 3. [*A Philosophical Essay on Probabilities,* trans. Frederick William Truscott and Frederick Lincoln Emory (1902)].

the objects of natural science is meaningless for the living subjects of science. They are able to do research, to be puzzled by "problems", to wonder, because they are driven towards a future goal that lies beyond their personal physical existence. Science is possible because man knows that his body is bound to die. The most important fact that we know of, every individual's physical death, is not a fact of the past or of the present but of the future. It has been said rightly that the root of all our knowledge is to be found in this prescience because it forces upon man the distinction between that part of him which is bound to pass away and those other elements of his existence which are not finished by this future event. "The future is the basis of our present evaluations," exclaimed the rediscoverer of the future, and its logical function, Friedrich Nietzsche.[12] It is, of course, an insight that has always operated; however, natural science, by looking backward on recurrent processes of the past, found no motive to mention this law of subjects. And our times, saturated with natural science as they are, ruin the very conditions of a prosperous natural science by carrying over to the subjects the rules that apply to objects only.[13]

The misunderstanding about the dependence of science on the power exerted by the future, and the pressure brought to bear upon men by our prescience of death, is a very serious one because it deprives the scientists of their dignity. On the other hand, it must be admitted that there is one particular reason why science in process should put aside this relation between the future and its actual operations. We don't know the future in the same way we know the facts of science. We know all facts of science because we know that we must die. Our belief in this future event is the basis of our scientific work in the field of matter. But we never must mix this belief with our method of research. Science is perverted if any rational concept of this future event would enter our thinking. When we die, where we die, all specific fears and hopes about the material realization of the future, must be kept out of our speculations. No scientific thought must be stained by speculations upon the material shape the future might

12. Friedrich Nietzsche, *Werke* XVI, 359.

13. Some remarks that point in our direction, may be found in William Stern, *Allgemeine Psychologie* [*Personalistscher Grundlage*] (Haag, 1935), pp. 386f., 551, and, with special application to the method of science, p. 77of. The principle is stated in Rosenstock, *Soziologie, I. Die Kräfte der Gemeinschaft* (Berlin, 1925), and in [Rosenstock], *Angewandte Seelenkunde* (Darmstadt, 1924).

show. Otherwise, prejudice, predilection, fear or hope would bias the scientific experiment. In this sense, the process of science is of that divine integrity of which Shakespeare speaks in *Troilus and Cressida* (IV, 5). Here the Greek king bids welcome to Hector, his enemy, for half an hour of complete armistice and mutual enjoyment. He praises the divine integrity of the extant moment in terms that sound as though they recall the happiness which we relish whenever we are steeped into the freedom and solitude of scientific research.

> "What's past and what's to come is strew'd with husks
> And formless ruin of oblivion;
> But in this extant moment, faith and troth,
> Strain'd purely from all hollow bias-drawing,
> Bids thee, with most divine integrity,
> From heart of very heart, great Hector, welcome."

Lest we misinterpret this welcome given to Hector by Agamemnon in a breathing spell between two battles, it begins with the significant pair of future and past: "What's past and what's to come". This should put us on the right track. It is from just this fact that both future and past are put aside for a moment that the interval which we call presence and which Shakespeare more rightly calls "the extant moment" draws its thrill. Science is the sublime freedom of man to surrender to his astonishment about the laws of life in face of the fact that his physical death is rapidly approaching and that the past is unalterable. It would be strange indeed, if this place of the scientific effort as a half way house between journey's end and journey's beginning had escaped notice among the scientists. Therefore, we need not be surprised that the first clear statement of scientific method is quite outspoken in this respect. In a famous passage, René Descartes tells us that he considered himself to be placed in three simultaneous domiciles, patiently recognizing his loyalties to the social past, fervidly believing in a final solution of nature's secrets and in the meantime consecrated to the pursuit of scientific doubt. Here we have the half way house of the scientific laboratory, of the scientific mind in the midst of its campaign. We may say then that Faraday and Descartes are in complete agreement as to the three tenses into which human time must be divided.

Any present time is created by a reaction of our faith in the future upon our loyalties towards the past. The presence is that portion of our life that we by our feeling certain about the future, can wrestle from the repetitive and recurrent part of our system, that portion won away from the laws of gravity so that we become free to grow, to add, to be changed. The present tense is a delicate product of a struggle between the pull from the future and the push from the past. The pull from the future is represented within a group or an individual by their beliefs. The push from the past is represented within their mind by consciousness and knowledge of facts. We said at the beginning of our investigation that the grammar of a scientist should lead to an understanding of the three dimensions of time. By an analysis of Faraday's grammar, and that is to say by an investigation carried out in a great center of the scientific process itself, one old long forgotten truth is re-established that mankind's future and mankind's past both precede its present tense logically. What we call present, is a result of the struggle between future and past. A mechanism has no future and therefore no presence. It exists as a repetition of the past. All mere recurrence belongs to the past. Science itself is not repetitive. The mind itself is alive; that means, it does not belong to the merely recurrent processes. Faraday expressed this, in his own language, but with great force when he said: "Electricity is often called wonderful, beautiful. But it is so only in common with the other forces of nature. The beauty of electricity or of any other force is not that the power is mysterious and unexpected but that it is under law, and that the taught intellect can even now govern it largely. The human mind is placed above, and not beneath it, and it is in such a point of view that the mental education afforded by science is rendered super-eminent in dignity."[14] This term "Law" is pointing to the recurrent past, "above" is Faraday's term for our "alive". Man, being alive, is suspended between future and past. He is able to create a presence, as an intermediary stage of transformation between believed destiny and innate fate. The present tense is a state of tense pressure between destiny and nature, finality and causation. Any one scientist fills this state with his doubts, his transforming ideas lest the ends that attract us from our goal, be missed by too narrow and too casual Causation. Any "error", any "preposterous

14. Silvanus P. Thompson, *Michael Faraday: His Life and Work* (New York, 1898), 185.

judgment", indeed, is endangering the fullness of our life, because it narrows the accessible means for our ends.

By discovering wider and deeper causes science eliminates unnecessary defeat and retreat. It is able to predict the equations of force and matter which supply us with the means for life. However, these predications have nothing to do with the "future" of civilization, the destiny of mankind, the goal of creation. Science only predicts the encroachments of all *lawful* processes upon his future. It can't wish to predict our future since that would deny its own vital importance. Michael Faraday's contribution to our knowledge is just that unknown quantity which makes prediction of the full future impossible, and science would defeat its own ends if it undertook to predict what difference its own achievements will make to society. "Faraday believed the human heart to be swayed by a power to which science or logic opened no approach."[15] Naturally, he must hold this belief. For science originated when modern man put his heart into settling in the present in the form of organized and cooperative doubt. He hereby tried to keep the vital balance between the believed future and the known past by enlarging the past and all its predictable processes infinitely.

Part II. The Founder of the Science of Life: The Tripartition in the Life of Theophrastus Paracelsus von Hohenheim (1493–1541)

HUMANISM VERSUS NATURAL SCIENCE

The times from 1450 to 1550 saw the rise of humanism and a new outbreak of fervor for the study of the classics. This interest in Rome and Greece did not discriminate against the pagan elements in classic civilization. The humanist fell in love with Homer and Virgil for their own beauty, with the city state and the empire for their own strength, without accusing the great heroes of the ancient world for not having been orthodox Christians.

The way in which this wave of humanists made its inroad into the Christian schools that after all, depended on the church at that time, was simple. The Bible existing in Greek and in Hebrew, the Digest of

---

15. John Tyndall, *Faraday as a Discoverer* (London, 1870), 185.

Roman Law in Latin, and the great physicians Galenus, Celsus, and Hippocrates, as well as Plato and Aristotle and Seneca and Cicero in Greek and Latin, the general war-cry was raised: *ad fontes*, back to the original sources. Translations, commentaries, anthologies, summaries, were swept aside. Read St. Paul, not St. Thomas Aquinas, read the emperor Justinian not the gloss of his thirteenth century commentator Accursius, read the Hebrew text of the psalms, not the latin Vulgate old, as it may be—these were the demands of an era of grammatical, linguistic and manuscriptual purism.

Luther's success in opposing Rome was intimately connected with his retrieving the Bible according to humanistic principles. For the new method approached the oldest layer of our traditions immediately everywhere. 1000 years of oral tradition and slow growth were pushed aside as a blind avenue that had led away from the classical texts. The New Testament after having been for a thousand years the first and oldest text of all the texts copied and edited and studied, now became the youngest book of the ancient world, or at least the contemporary of Plato, Caesar and Virgil. Any text was welcome provided it was old. The movement did not halt at any special field, mathematics, or botany. The Greek Theophrast became once more, in his Greek garb, the authority he had been in the Middle Ages, for plants, Ptolemaios for geography.

It is easy to see why progress in the understanding of the classic texts was a draw-back in the field of natural science. Here, it repressed the drive for direct observation and experiment. Since Celsus taught that four humores (saps) filled the inner part of the human body and produced, by their divers concoction, the four tempers of human character, it seemed superfluous to begin the study of life all over again. No wonder, then, that the first real scientists were persecuted and scorned not by the church but by the humanists. The humanists being as eager as any Don at Oxford today for writing and speaking an immaculate Latin, found fault with any member of the profession who neglected these formal arts and preferred the soiling contact of reality to elegance of style. The nearer a scientist came to the problems of biology and psychology, the more difficult became his position. Anatomy, Mechanics, and Astronomy were less imperiled than research in the field of living substances where many more ancient prejudices survived. The tradition of Humanism in later centuries was generous toward the anatomical findings of Leonardo da Vinci or Vesalius, towards Kopernicus and Galilei. It never pardoned the pioneers in the field of life. Almost, one is tempted to say that the humanists living in a world of printed books, of dead paper, little objected to additional discoveries in other departments of dead matter, such as stars or stones. They turned a deaf ear to all attempts at rediscovering the secrets of life. The result of this humanistic prejudice was a colossal predominance of mechanics in the last four centuries. The study of the dead by far prevails the study of living substance even today.

Humanism, in order to defend its birthright of dealing with classical texts, did not shrink from lies and legends to monopolize fame. Modern textbooks on the history of science are filled with these distortions of facts, the arbitrary choice of heroes, the suppression of real pioneers.

The most illustrious victim of Humanism was Paracelsus, the founder of modern biology, biochemistry, pharmacology and internal medicine. His fatal conflict with the Humanists of his time is of so gigantic dimensions that modern scientists like Dampier who try to give a history of science, remain helpless before this tragedy. Since they are convinced, at the start, that Humanism was progressive, they cannot conceive of any falsehood in the official record.

Still, the sources are all available now and were available when Dampier wrote in 1927.[16] They disclose an example of the reckless terror practised by any victorious body of thought, and of the torments of real genius that has validity for all times. Fascinating as the life of Theophrastus of Hohenheim undoubtedly is, its greatest importance lies in suggesting the true relations between books, life, progress, and legend.

The Antecedents of a New Life

In 1493, Paracelsus was born as the son of a physician in Eastern Switzerland. He was ten years younger than Martin Luther. His birthplace, the village of Einsiedeln, is a center of pilgrimage famous even in our days, because of a miraculous Madonna in the Benedictine Abbey. One might compare such a center at the end of the Middle Ages to

---

16. *Ed.* Probably a reference to William Cecil Dampier, *A History of Science and Its Relations with Philosophy and Religion*, which appeared in many editions.

a place like Saratoga Springs or Karlsbad today, or to the temples of Asklepios on Rhodus in antiquity, instead of comparing them to modern Lourdes. For no dividing line then separated medical treatment and religious elation. This division to which we are so devoted today, between physician and priest, did not mean very much.

The father, William of Hohenheim, was a great admirer of botany, and the surroundings of Einsiedeln offer a rare opportunity of finding the alpine flora in all its wealth. "On the meadows, banks and in the woods by the Sihlstream and in the valley where swamps abound, spring, summer, autumn, and winter bring countless plants to bloom and fruition. In the meadows, primulas, gentians, daisies, salvia, ranunculus, orchids, camomile, colchicum, borage, angelica, fennel, kuemmel, poppies and martagon lillies succeed each other. In the woods, pirolas of five varieties, woodroot, belladonna, datura, violets and wild berries are plentiful. On the banks and roadsides are campanulas, foxgloves, chicory, centaurea, many different veronicas, plums, mint, thyme, vervain, smilax lychnis, St. John's wort, potentillas, ribwort, and witch herb; on the swamps are the mealy primrose in great patches of lavender and purple, sundews, myosotis, pinguiculas mallows, equisetums, selaginella, a rare orchid, and on the moors and mountain slopes erica, azalia, alpenrose, saxifraga, grass of Parnassus, dianthus, wild plum and wild berries again. And these are but a few out of a much larger list" (compare the book of Miss Stoddart, p. 25f.)[17] All this surrounded the child and in honor of this environment he was baptised Theophrastus, with the name of the greatest botanist of antiquity.

This vegetation blossomed around the sanctuary of a medieval clerical community where healing, prayers, immemorial wisdom of the universal church and local therapeutical experience all were blended together.

From this first environment Theophrastus was transferred into a second of no less extreme character. After the quintessence of medieval civilization the most modern and radical industrialism. William of Hohenheim moved to Villach in Karinthia as a teacher of chemistry and a physician. Here the wealthy Fuggers, the Pierpont Morgans of the times, operated mines. These mines, Paracelsus described gladly: "At Bleiberg is a wonderful load-ore that provides Germany,

17. Anna M. Stoddart, *The Life of Paracelsus . . . 1497–1541* (London, 1915)

Pannonia, Turkey, and Italy with lead; at Huettenberg iron-ore full of specially fine steel and much alaun ore, also vitriol are of strong degree; gold ore at St. Paternion; also zinc ore, a very rare metal not found elsewhere in Europe, rarer than the others; excellent cinnabor ore which is not without quicksilver, and others of value . . . and so the mountains of Karinthia are like a strong box which when opened with a key they reveal their treasures." At Villach, the technique of production anticipated by centuries the capitalistic and industrial forms of our days. Big water pumps, complicated screwing machines, and elaborate processes of chemical precipitation were in use. Even the forms of business were radically modern. For the need for capital had forced upon the mining industry the system of Kuxs, i.e. of anonymous ownership of stockholders. Furthermore, the miners were working in shifts, regardless of sun or moon. And a free market for the goods sent over Eastern Europe, and for the labor supply that took in capable men of all nationalities, rounded off the picture of an absolutely modern society.

These two phases left Theophrastus von Hohenheim with all the existing information that the Middle Ages and Modernity were capable of offering. As a matter of course, he learnt Latin, and wrote a terse and precise Latin ever after.

The particular constellation of his antecedents did not end here. His early life contains a third chapter. And here, the great new principle emerged, a new kind of intellectual intuition developed, combining the rare advantages of his education with the impressions of the Wanderjahre that now followed.

Today, we are accustomed to assume that a man who talks about mushrooms will look up the places where mushrooms grow. On the other hand it is widely known of the 16th century that students would travel and rove between the seats of higher learning. So when we hear now of a period of travel in the life of our hero, we may think of it as very natural. It would be a mistake to deal with this third chapter in a casual way. Between 1514 and 1526, Paracelsus travelled through every country of Europe and the Near East. He was in Venice and Stockholm, in Spain and Greece, in England and Denmark, and of course, in France, sometimes studying at a university, mostly, however, earning his way as a physician in one of the armies during the many campaigns of those years. Now, in at least a dozen passages, Theophrastus deals at great length with the significance of his journeys. When we recall that Columbus asserted to himself and others that he went on a Crusade for converting the heathen, and that most of the wandering scholars begged alms on their way from one "Generale Studium" (i.e., University) to the other, we will understand the following quotations to be highly original.

> "That I travelled widely and resigned me to a life of migration without any stable place or home, is reproached me by the regular scholar. They say that I am less useful. But you will understand my resenting to see a merit turned into a crime. At home nobody learns his craft nor will he get wisdom from behind the furnace. For the arts are not enshrined in our native place. They are not given to one man only nor to one place only. We have to collect them from different places and to get them there where they blossom best. In this my witness shall be the firmament of stars, where the inclinations are peculiarly distributed, not compiled in one section, but according to the nature of the upper sphere the rays expand everywhere. Is it not righteous, then, to persecute their ends and to observe them in all their filiations? Science is not obtruding itself. We have to look out for it. Hence I was obliged to go in search of science since she would not come by herself. When we wish to be with God it is we who have to go to him who speaks: Come to me. When we wish to learn about a person, a city, a country, or the nature of the sky, or the qualities of an element, we must betake ourselves to that place. Could anyone become a good geographer behind his stove?
>
> "The same is true in medicine. For the diseases themselves occasionally migrate and tour the whole world. If a man wishes to know as much as possible about disease, let him migrate. English humores are not Hungarian, and Humores of Naples differ from Prussian. And so you go and look them up each in its place. Wisdom is god's gift. Where god puts the gift there we have to seek it. This knowledge is arduous once we perceive that we must seek god's gifts where they are hidden and that we are in a way compelled to go where he placed them. It is not comfortable and it is no fun to undergo all the hardships of travel. To sit at home in your mother's lap and to eat your fine meals, to have the temperature, hot or cool, in your home to your pleasure, and to wear the clothes you like best certainly is

the easier proposition. But to travel is the only way of reading the book of nature. That I can prove from its very character. A man who wants to explore nature, can't help turning the volumes of his library page by page with his feet. Scripture is deciphered through letters. Nature is explored through our feet taking us from place to place. As many countries in the world, as many pages of nature. Such is nature's manuscript and that is the method of turning its pages."

In another work, he describes the character of tartar, gallstone, calcoli, all the various solid concretions formed in any part of the human body in excess. (In fact, Theophrastus was the first to give a unified interpretation to these various sediments or "scoria" formed by the process of living. This discovery enabled him to reject the Galenian theory of the four humores that blinded medicine for another two hundred years against the formative powers in physiology.) In fighting the classical tradition he exclaimed: "Man's calculous disease should be judged with regard to the stones or calcoli that are prevalent among the people of a particular country. And since no variety of calcoli is described in the libraries containing theoretical books we ought to seek another library in which the story is told by demonstration and which contains the species genuinely and completely. The universe is this library of which not one part only may be read, but we must keep volitant through all its elements, and through its upper and lower sphere."

"The human mind knows nothing of the nature of things from inward meditation. I have to remind you that fantastic imagination does not adorn the physician. That which his eyes see and his hands touch, that is his teacher. Take an example: A denizen of a monastery who lived in it forever and saw nothing except monastic habits and ritual, will find himself ignorant of any other habits when ever they occur. Put before this man the problems of the calcoli. He may try to decide them by monastic speculation which springs from mere human imagination. This monk never can read the true fundamentals. Still he can be the best expert for his monastic rules. And this applies to the situation in medicine. We have doctors indulging in speculation and bookreading and they will not hear much more than the monk who listens to the chiming of the bells of his chapel. I mention all this in order that I may explain why

I eliminate and reject the description of calcoli given by the ancients. We ought not to hand down speculative knowledge, and we should use true demonstration.

"And this is not restricted to the doctrine of calcoli. It is required in universal medical theory. For this theory takes its foundations from the things existing in earth and water. So from these two elements all description of tartar will start. I do not establish any limits to this library. Paper with the wrong theory on its sheets may tell us nothing about the true origin. Whereas earth and water, `i.e., the material world, are like true matrices and are the genuine books and manuscripts. Since this is the kind of book for medicine, I confess that where I am going to stop a second, a third, a fourth volume will begin until all are conceived and accomplished. In other words, if I claimed to have travelled through Asia or Africa and to have turned their pages that would not be true. Through the larger part of Europe I went and explored. Who could penetrate into all the corners of the world singlehanded? I am writing for Europe; and I doubt if my writings are profitable for Asia or Africa. As every day has its affliction so every place is close to a special evil. This applies to every nation, province, valley or climate. Hence every one of us is like a cosmographer or geographer, turning some pages of his art with his feet and surveying with his eyes the characteristic element of each place. In this way, we shall build up a survey of all countries so that we may learn how many species in each single country exist. Let every physician collect all data about the peculiarities of his district. When this is done in the same spirit by all doctors of all places and countries, then finally the book of medicine such as it consists in (the material world as represented by) earth and water, could be written out, with a sure foundation, in paper and ink, and be sealed. Even then, this book of paper would not be more than a map of the real world. No map can be read by a man who never saw the real world. However, you may attain the true library of the physician's religion once you amassed your erudition out of the genuine book of nature and tested all your work by this true touchstone of philosophy."

As far as I can see, we here have the first and most glorious program of the things that came to pass in the four hundred years that followed, in terse and pithy language. It belongs, one should imagine, to any history of science that is true to the principles laid down by Paracelsus and accepted by all great scientists as their method of research. Unfortunately, these principles are not applied when scientists write the history of their science. This is not the place to give the interesting causes for this failure. Our quotation gives the lie to the humanistic legend that Theophrastus was "an arrogant quack as ever one lived", as a recent expert on Humanism thought fit to call him. I question Mr. Preserved Smith, the author of the *Age of the Reformation* [1920], and shall question him again later, as I question Mr. Dampier, if they were acquainted with any of the central works of our man, when writing about him.

Theophrastus transferred his new method from mere geography to the more general field of space when, in his main book on philosophy, he uses these words: "Open your eyes. Do not stare at the stars only and their revolutions as the astrologers do. Observe what is at your feet, also. A man should not turn his eyes to one place only, but let them roam through different spaces and places. Besides, open your ears and try to listen to unheard noises. Nowhere is earth so empty or inane that eyes and ears should not have many things everywhere to look at and listen to. The further your legs take you to foreign countries, the more your eyes see and your ears catch. Everywhere you fall into the midst of god's works and miracles the inspection of which will polish and illuminate you. Neither do all animals stay in one spot nor are all fruits collected in one garden. We must keep roving through all the kinds of the creatures so that we come to know them, as the expression is, 'from beginning to end'. God works in heaven and earth, fire and air what he wills. There you turn and spot him where he is at work."

We see that travelling was only the most obvious and external form of the new method. Since any space contained its own text, the infinitesimally small space of a nut shell deserved laborious investigation. Experimentation in the laboratory, then, is pressed by Theophrastus, much in opposition to the ivory tower tradition of the Galenians. He says:

> "The analysis of those material things that grow from the earth and are easily combustible such as all fruits, herbs, flowers, leaves, grass, roots, woods, etc., is carried out in many ways. By distillation first watery distillate is extracted, then

the gaseous products, the third product is resin, the fourth the combustible, the fifth the ashes. When this analysis is performed, many splendid and powerful remedies for internal and external use are extracted. Therefore I praise the chemical physicians. For they do not resort to loafing or going about gorgeously in satins, silks, and velvets, with gold rings on their fingers, silver daggers at their hips, white gloves on their hands, instead, they tend their work at the fire patiently day and night. They do not go promenading, and find recreation in the laboratory wearing plain leathern dress and aprons of hide on which to wipe their hands, thrusting their fingers amongst the coals into dirt and rubbish and not into golden rings. They are sooty and dirty like blacksmith and charcoal-burner; hence they are not showy, waste no words, don't gossip with their patients, and do not advertise their own drugs. Too well do they know that the master is known by his work, not the work by its master. They are convinced that talking and chattering do not help the sick nor cure them. Therefore they leave those things alone and busy themselves with their furnaces and learning the processes of chemistry. And here is a list of these processes: distillation, solution, putrefaction, extraction, calcination, reverberation, sublimation, fixation, separation, reduction, coagulation, tinction." (From his "*de natura rerum*").

In fact, we owe to Paracelsus the notion of gaseous processes and the terms "reduction" and "reduce" which today are so familiar. Theophrastus, it is true, did not say "gas"; his first successful disciple, van Helmont, put this abbreviation of the word "chaos" used by Theophrastus in circulation.[18] However, the concept is wholly Hohenheim's. And like his discovery of the unity in all sedimentation, "gas" put an end to the division between air and solid bodies, by definitely proposing the idea of different aggregate status for any substance.

Of this ardent observer of real processes Mr. Preserved Smith writes: "He worked out his system a priori from a fantastic postulate."[19] So deeply rooted is the a priori in the mind of the biographer of the King of the Humanists, of Ersamus of Rotterdam, that an adversary of Humanism is not treated with the fairness accorded to any criminal:

18. *Ed.* Jan Baptiste van Helmont (1580–1644), Belgian chemist.
19. *The Age of the Reformation*, p. 513.

to be judged by his own acts and words, and not by the slander of his humanistic enemies. The outcome is this fantastic line of criticism.

The new method and the new vocabulary implied new social conditions too. The academic clan never tired of scolding Hohenheim for living with the common man. He roomed in the pubs and taverns like a hedge lawyer, and he talked mostly to peasants. "You are telling me that I should converse with the doctors at Louvain, Paris, Vienna, Ingolstadt, Cologne where I had real personalities under my eyes, no peasants, no tradesmen, aye, masters of theology or medicine. However, I inquired and searched for the true and experienced arts of medicine not from doctors only, also from barbers, surgeons (at that time kept far apart from physicians), learned physicians, women, practitioners of sorcery, alchemists, in cloisters, from the noble and the commoner, from the wise and the simple."

Accordingly, among the seven rules for the surgeon in his Antimedicus, the first runs as follows: "He shall not consider himself competent to cure in all cases. Assume the case of a doctor as wise and learned as is conceivable, there will come an hour where a case puts to shame all books and all experiences and startles him by its unfamiliarity so that however learned he may be, with regard to it, he feels lost. For that reason, you and I should learn daily, note, observe diligently, despise no communication, nor trust ourselves too much, and above all, realize how little we can do, even although a doctor and a master. Therefore, you have got to remain in the state of learning because who is able to do everything or who can know where all cures are to be found? You must travel and accept without scorn that [which] comes to your hand."

No wonder that this "arrogant quack", with the "fantastic a priori postulate", wrote, with the humility of genius: "Now it is perfectly true that earth contains many things which I do not know and which are unknown to others also. For I know very well that God is going to disclose many remarkable things of which we all never knew a bit. And it is true also: nothing is hidden which is not to be disclosed. Hence there will come, after me, one whose great power is not yet in existence, and he will disclose it."[20]

So we see the maturity of a man who stands equally far off from snobbery, from bookishness, from self-complacency. He had been

20. *Werke* III, 46.

raised in the arms of all the gods of reality, at Einsiedeln, at Villach and in his Wanderjahre; the living spirits talked to him whose very existence the snobbish, bookish and self-conceited man of society ignores or tries to ignore.

And at this moment, in the 33rd year of his life, A. D. 1526, this man was called to fill a chair at the citadel of humanism, at the recently established University of Basel. A cure that restored the health of the illustrious printer Froben, [who was] a friend of the great Erasmus and [of] the political leaders of the city, paved the road to the unusual appointment.[21]

For one moment, both humanism and science, might seem equally modern, equally valuable for the Reformation. Actually, two contradictory forms of thought, of research, of social standards and of faith clashed in the tragedy of which we have to speak now.

Theophrastus becomes Paracelsus

An unbroken chain of authorities waited to receive the student of an academic community. From the very beginning, he had to sacrifice his worldly ways of thought and speech since all teaching was done in Latin. Any new language implies a new loyalty. This loyalty's meaning was made conscious when a student graduated. He would take an oath "On the book opened and on the book closed," promising to read first and to comment later. No teacher was allowed to teach his own doctrine; down to 1800, and in many colleges even today, teachers lecture "on" certain books written by others. Nowadays, however, the book may be written by a contemporary. This was not done as late as the days of Kant. The great Immanuel Kant had to lecture on an older man's book during all his academic career.

Into this world of Latin lectures "on" authorities, Theophrastus jumped. His proud name was exchanged for a new one, Paracelsus. In his days, any Tailor became a "sartor", Descartes became Cartesius, any number of Latin translations of French, German and English names could be listed. Theophrastus of Hohenheim continued to call himself by his original name; alas, posterity only knows him by his nickname, a name given him by his enemies as a stigma on his academic career.

21. *Ed.* Johann Froben (1460–1527).

And so everybody speaks of Paracelsus today; I myself am compelled by the weight of tradition to call him that; and yet, I know that the man himself would feel hurt if he heard me quote his enemies. They seem to have taken advantage of his frequent use of the Greek preposition *para* (*praeter*), which two titles of [his] books, the Paragranum and the Paramirum, contain. So Paracelsus literally means the "Super-Celsus"; Celsus was a twin of Galenus, the old Roman medical man.

The originality of Theophrast, his aloofness to bookish tradition spurred the academic clan to hail him ironically as the man who made Celsus superfluous, being a kind of Super Celsus. Like an I N R I on a cross, the nickname Paracelsus stuck. It greeted him probably inofficially in his very first days in Basel; and as late as 1800, his disciples would call themselves Paracelsists. And Theophrastus himself, after some reluctance, acquiesced at being called this by others. Since the thing is apparently unknown, I wish to say that he avoided the name for himself in all formal publications and statements till to the end of his life.

This Super- physician who as we saw before simply kept his feet on real mother earth, his mind on real data, delivered his lectures in plain German. No modern language, in 1500, was trained as a vessel of science. Paracelsus created a terse and simple German style for all he had to say and tried to express. Since it is hardly possible to prove this to anybody unable to read him in his native tongue, the tremendous influence of his creative style, and its enchanting power, again, only may be pointed out by the fury of his enemies who, from his second German name, Bombastus (which means: twig of a tree), deduced the word "bombastic". In fact, this word for a swollen and exaggerated style is derived from a Greek term for silk, cotton; and the bony and pithy sentences of Paracelsus are everything but cotton. The victim of Humanism, of course, had to pay the price for being not interested in the rules of Golden Latinity (which, by the way, he wrote as easy and as fluently as any one of them; which he yet did not think very important). The first man in the Western World for centuries who lectured in a University openly in his native tongue, was stigmatized not for his simplicity as you might expect but for his "bombasticity".

When we turn to the content of these lectures in the native tongue, we may get our information from his program. Any professor would publish a program of his lectures in those times. In this printed sheet, a list of his authorities, his "assigned readings", so to speak, was given, and the method of his course on these authorities. Theophrastus certainly would have been insincere had he advised his students to read the accepted books. He announced his authorities in these terms:

"Smooth talk in different languages does not make a physician nor the reading of many books; he is made by the knowledge of the material world and its hidden powers. The physician's business is to know the varieties of the processes that take place in the body, and the right remedies that exist in nature, with insight and industry. I, therefore, am going to dictate to you books of which I am the author myself—"*quorum et ipse auctor*", for the program, at least was edited in Latin—and which are based on a long and trying courtship of Lady Experience. In case you are willing to be led by me into these new paths of study, come hither to Basel. However, it is not until you have listened to Theophrastus that you may understand and judge his purpose."

Today we know that this was the program not of a professor appointed by a city, but of a genius appointed by god to bring forth a complete system of biology, chemistry, physiology, including new drugs, the greatest connoisseurship of all mineral waters and medical springs, a clear insight into the reforms of pharmacies, hospitals, and the whole practice of hygiene, miraculous healing etc. This man understood—what is rediscovered in our generation—that mental diseases and psychic wear are two quite different classes of illness. He was an experienced surgeon, too, and strangely enough, here, the envy and hatred of his fellow physicians stopped because this was outside their domain. What I was stressing, by this enumeration, is this: This man's program was modest in comparison to the wealth that was taking shape in this mind during the 34th year of his life. It sounded impossible and arrogant in the halls of tradition. The man, who exalted German for the first time to the range of a spoken scientific and yet pithy language, stated in a few sentences that a life unique in its antecedents and its opportunities was ready to bear fruit in others. In the sudden contact between his new universe of experience and the old requirements of lectures and bluebooks, all his glimpses and insights and his many draughts and designs crystallized. Of course, in the year itself, only particles of this new treasure could be precipitated. Nevertheless, it is a fact that this one year marks an epoch in the whole rich production of Theophrastus. It seems as if every sentence spoken at Basel, every question put to him in these few months, every idea articulated under the pressure of regular teaching here, was, by its belonging to this extraordinary year, indelible, for ever asking to be further developed. Like the promises which an honest man makes good, these words were all followed up by weighty and voluminous works.

Rarely may we observe this process of crystallization that is hidden behind the stereotyped term of character. The Greek word "Character" means face of a coin. Now, a man has no character as a child. Character befalls us in our first full exposure to the world. And in the life of real genius, this process leads to a real change of the universe because something new that never existed before is produced by life's melting pot. Theophrastus acquired his character *indelibilis*, his appointment by God, through his conflicts with the men among whom he had to live by the odd appointment to a professorship.

Scarcely was he appointed when the faculty objected to his academic training and credentials. He applied to the City Council that upheld his good and valid title of doctor, acquired at Padua in Italy a decade earlier. Soon the attacks became more serious. Theophrastus, according to his appointment, was supervisor of the pharmacies of the town. He soon found out that a racket was in existence between pharmacists and physicians, and putting an end to this exploitation of the sick he naturally incurred the lasting hatred of a craft he had exposed. The financial background of the passions roused against him is of peculiar interest; humanism always was inclined to ally itself with the rich and, for that reason, was easily bribed. Theophrastus was incorruptible, Erasmus was not.

Against this background of suspicion and fears, the teaching itself was polluted by the interference of outsiders. His students were bribed by his colleagues to report his lectures so that they might get hold of some material against him. His programatic sentence *"quorum et ipse auctor"* was too deeply resented by the herd of ruminators; and well were they protected by law. We find thirty years later a man in England, a medical doctor, forced to ask official pardon for having criticized the writings of Galenus! No wonder that an enraged colleague of Theophrast quoted two decades later, as proof of the man's madness, the *"quorum et ipse auctor"*.

Very soon the attacks on him took the form of songs and pamphlets. In his examinations, his colleagues intervened with insulting questions. His nickname "Paracelsus", Super-Celsus, was invented now and passed around. In one poem, Galenus himself returned from Hades to scold the innovator. What did the students do? This raises an interesting issue. We are brought up under the tacit understanding that youth is magnanimous, generous and full of sympathy with genius. This is a half- truth only. We may say: sometimes, some young people are this way. The majority never is, at least it was not in the great case before us. Perhaps, to understand this situation, we need a parallel from light. Sun-rays and the light of distant stars travel quite a bit before they reach us. It is hard for us to grasp the fact that the light we see today generated a full century before. Is it not true that the light generated by a human heart undergoes similar laws of irradiation? It is not true that a man sharing the same room with genius and listening to his speech, is able to get at his thought. When a new light shines up among men, in its first year of appearance it hardly is visible. As far as it is observed at all as "something", we may be sure that it will be misplaced and disqualified and classified under "madness", or "heresy", or ridiculed. The thoughts of man travel as slowly from one man to another as the light of the stars.

For that reason, it is not to be wondered that the students of a new humanistic center were not prepared to understand the new deity of experience and experimentation and her prophet, Theophrastus von Hohenheim.

Hohenheim, himself, loved to instruct and to impart his knowledge; and in every country, apprentices and fellows gathered around him as assistants so that they might learn some of his secrets. Once he talked about them frankly: "So many disciples are conceived by righteous physicians and yet they turn out failures not following their preceptors. Coordinating teacher and pupil is impossible except both remain in the realm of immediate experience. Though I begot physicians by the hundreds, only two were a success from Hungary, three from Poland, two Saxons, one from Slovenia, one Bohemian, one Dutch, not one Swabian [this last remarks, perhaps, includes his years in Strassburg, Colmar, and Basel]; yet, of each race they had been numerous. Each saddle my teaching after his whim. One abused it for his purse; another for his vanity; again another emendated it. Some thought themselves more intelligent than they were. Some were able practitioners however, without the subtle understanding; some were clever, but, being clumsy, they became arch scoundrels."[22]

On St. John's Eve, the 24th, of June, the students usually had great fires, and threw into these fires all kind of rubbish. Attacked and slandered from all sides, the "Super- Celsus" calmly approached the fire and threw one of the standard textbooks of medicine into the flames.

Fashionable Humanism, vested interests, and the anxiety of the students for a regular career, all three were roused by the personality that came from a world so far outside established society, from god's world itself. His great friend and protector Froben died when Theophrastus visited Zurich, the neighboring place to gain support there from his colleague supervisor.

The death of Froben brought things to an end. Without Froben, he was unable to print his new ideas. Without winning the general public for his ideas, certainly he could not secure the toleration of the local powers. Before retreating, Theophrastus tried, very soberly, to rally with his colleague of Zurich for a publication on drugs, the supervision of which was their common duty. Since humanistic enmity, for centuries, decries him as a drunkard, crazy, a quack, overbearing, impossible to live with, it only is fair that we should investigate one single occurrence where we are able to see him act within society. His colleague, Klauser, had introduced him to the students during his visit. Theophrastus, in a letter, reminds him of their pleasant gatherings and revives them by some merry remarks. Then, he goes on to describe his lecture on the grading of prescriptions, a topic of first rate importance for both correspondents in their character as city-physician. And since he notoriously rebukes Galenus, he entrenches himself behind the authority of Galen's greater Greek predecessor, Hippocrates. Shielding his own revival of research by Hippocrates he releases the same forces that every innovation within the last humanistic centuries would set in motion. Adding weight to the Greek part of the classics always was used as a means to develop a new aspect of reality away from the Roman tradition. At the end of his book, he asked for help to print the book on grading.

The letter is a specimen of good manners, good style, and superior sobriety. The manuscript, however, never was printed in Paracelsus'

22. *Werke* VI, 55.

life-time. The lack of urbanity, then, is not the reason of his failure. What, then, was it?

I venture to think that we may peer into the machinery of hostile reactions by studying his way of quoting Hippocrates. Humanism was based on written authorities, on the undoubted existence of "classics". When the Humanists reprinted Hippocrates, they cackled because a new Greek author was added to the list of classics. The classroom needs authorities. In the Chair, we all are compelled to rally behind big names, great books, established reputations. Reluctantly or not, any public teacher has got to quote other books and other authority. This seems to be a kind of sociological law. The law probably is caused by the unreal character of a classroom in which we assemble, three times a week, for mental rumination. Public teaching is impossible without pointing to events that are more direct, more real, more adventurous than the processes in the classroom itself. It seems that the environment in a class is too unreal to endow our words with the halo of complete reality. The deficit is made up by quotations from more direct, more fully real authorities.

Now, Theophrastus was accustomed to outdoor life. He never quoted others when his own experience supplied him with what to say. He wrote the proud line: "Never did I write a word without experience!" When entering public teaching, he suddenly felt the rules of the new game around him, and honestly made an effort to comply with them. He quoted Hippocrates. Unfortunately, he quoted him as a colleague who corroborated his own findings. Everybody could scent that Hippocrates was not quoted because he was a classic, but because he was right. We shall mention later how this invincible law of teaching determined Theophrast's own literary reputation. Others lack genius, Theophrast might have fed seven medical faculties with his inspiration. All his problem in Basel and in his after-life was authority, and once more authority. It was queer how he finally acquired it.

Physicians, colleagues of the faculty, pharmacists, students gathering against him, a rich patient of Hohenheim, a canon, thought that he well might abuse the lonely figure. He declined to pay him the promised fee after a successful cure. The insolence of the man being obvious, he was sued in court for the fee; when the judge, on the formal reason that the fee was not fixed by Theophrastus but, in the form of a public promise to anybody, by the canon himself, dismissed the case. Theophrastus used an expression of disgust. Now, he was caught. This was contempt of court. His enemies now could proceed without even mentioning professional motives. Exile into the higher Alps was proposed. Hohenheim left the city, his mission as the teacher of a new science now being perverted into a fight about his accidental "contempt of court". He escaped to a friend in Colmar where Gruenwald's altar stands.[23] His resume of the year was given in lapidary style by himself in a letter, *veritas parit odium*; Truth engenders hate.

After-Life

From his appearance at Basel to his death at Salzburg a sandy desert seems to extend itself. In Basel, his enemies were able to get rid of him under shallow pretexts which enabled them to evade the real issue. In Salzburg he succumbed to an assault against him by the hirelings of his lasting enemies, the humanistic physicians. Morally and physically, then, the profession tried to destroy him, and did destroy him in any worldly sense of the word completely. The years between 1527 and 1541 were one constant fight uphill. Sometimes he was turned away from a town because of his shabby clothes though sometimes, as in Pressburg, he was received like a prince of science. He wandered, between Rhine and Danube, in Switzerland and Austria, in Bavaria and the Tyrol. His reputation as a great physician did not abate; he was able to leave a considerable amount of money to the poor of Salzburg. And his generous attitude toward the poor survived as a tradition for centuries. As late as 1830, the people at Salzburg, during an epidemic of cholera, prayed at his grave for protection.

The desert, then, that surrounded him, was not external starvation so much as the permanent danger of complete oblivion, of seeing the "great monarchy of medicine" that he felt had come down in his times upon earth, and had crowned him first, without any presumption on his side—of seeing this unique revelation cancelled and deleted from the book of history as if it had never existed at all. It took him years to realize the full intensity of his outlawry. At first he assumed quite naturally that local powers at Basel could not represent the universe of science. He was ready to fight. He put forth his claim

23. *Ed.* The Isenheim Altarpiece by the painter Matthias Gruenwald (1470–1528).

in violent language breathing the excitement of a man who feels that he stands for something sacred, entrusted to him for all mankind. He soon found out that 'the powers that be' wherever they might be, identified themselves with their representatives at Basel. When it transpired that the City of Nuremberg was willing to print one of his manuscripts, the medical faculty of the University of Leipzig intervened and successfully prevented the print. In 1537, the Estates of Styria promised to print four tracts which he dedicated to them. They never kept their promise, and the manuscripts still rest in the Styrian archives. The only considerable work published in his life-time, was his surgeonry in which the medical profession was less directly interested. After forty-eight years of toil in life, Theophrast von Hohenheim had to wait another forty eight years after his death till his real medical work was given to the world. By a chain of accidents, this vindication of his work did not contain the surgeonry. This gap in the monumental edition of the faithful Huser, in 1589, is like an ironical remark of fate, whispering: Look here, the only thing his contemporaries were willing to tolerate is less important and valuable than the last sheet preserved by faithful friends in secrecy and appreciated by posterity as starting a new era in our attitude towards life.[24]

His difficulties were increased by his unwillingness to take sides in the religious struggles of his time. He was neither a Lutheran nor a Baptist, the two aggressive groups of the period, nor could he, with his great vision of a living universe, overlook the imitative character of ceremonies and miracles in the traditional sense. In not being able to identify himself with any one of the three parties, he forfeited the claim for support which even an unworthy member of a party gets. He had friends in all groups; and he was bound to disappoint them all in due course when they wanted him to join their religious fight. This would have meant high treason to his new "Monarchy of Medicine." Being the only general and soldier of this new realm, these fourteen years put upon him the burden of making an everlasting impression on a world that did everything to prevent him from leaving any vestige whatever.

24. *Ed.* The reference is to Johannes Huser (d. 1600?), the first editor of the works of Paracelsus. For details on this and subsequent editions see online the University of Zurich's "Paracelsus Project".

No monastery, with its great collection of manuscripts, could be of any use for Theophrast. Public archives and libraries did not exist. Of this dilemma, Hohenheim disposed by cultivating friendship.

Everywhere he made friends, one or two perhaps, but intimately affected by something inexpressibly great in the man. These people became the trustees of his knowledge and the manuscripts which he dictated. Without a place in which to stay, during constant medical practice and traveling thousands of miles, he managed to produce about ten thousand pages of manuscript in these fourteen years. This alone shows his great power of concentration, and the high degree of discipline; it suffices to put to shame, as also do his pictures, all the slander about his being drunk constantly, one of these silly inventions that have been used against any President of the United States, in the same way as against Paracelsus.

However, this was not all. His task was much more complex. We, in our luxury of libraries and books and scientific progress don't see easily the real obstacles of a great man in the sixteenth century. Leonardo da Vinci who often is called the greatest scientist of his time, left some unreadable notebooks and sketches behind him, a collection of hints, divinations, anticipations. To call such a man the greatest scientist, is the cheap apology of an epoch that is willing to concede everything to a great artist, but knows little about the sociology of knowledge, about the problem of placing knowledge in a place where it can bear fruit, where it can begin to change the daily practice of others, be taught in schools, transform the world with the power of a new gospel. This, and not some ideas scribbled in a notebook, is so difficult to achieve.

His real difficulty, therefore, went much deeper. In times where no *Encyclopaedia Britannica* gives the illusion to everybody of his having access to the whole universe, he stood for the comparative study of natural phenomena, for travel, for a map of the world based on the collaboration of thousands of scientists all over the world. He stood for a new place of biology and medicine in life and society, by which the meaningful character of illness, as a process of life itself, not as an external and accidental "thing" from outside, would become visible: "Man acts continually against the laws of his nature. A time will come when disease will be the result, because the organism requires a period of rest and a renewal of strength to expel the accumulated poisonous elements. If the physician attempts to prevent such an expulsion of poisonous elements, he attempts a crime against nature and may cause the death of his patient (On the Character of Poison)." He knew that mercury was the thing to use against syphilis, he knew of the therapeutic usefulness of zinc, and of laudanum as well, three new and important drugs. He knew how many diseases were of mental origin, and in an ingenious scheme saw man hanging, so to speak, in five systems of different range: the mechanical and physical, the neurologic or nervous, the psychic and mental, the spiritual or astral, and finally, of the god head. That is to say that a man can fall ill and practically falls ill on one of five levels. Sometimes he is wounded mechanically and physically; then he ought to be treated on this level. But any one of the more complex systems in which he is embedded also may make him ill, and give very similar symptoms. Curing him on a wrong level would not restore his health. A physical disease, originating in a mental condition, a nervous disease provoked by psychic wear, a mental disorder caused by a "social", this has become our term for "astral" lesion, depend for healing on first being traced to their source.

The physical level was real and deserved most intimate experimentation; but it did not mean what we call physics, the science of the dead parts of our physique. Body did not mean a carcass; the natural world did not mean a collection of glowing stones. Theophrastus wished to follow the light of nature because nature was to him the living cosmos. He used nature in the sense A[lfred] N[orth] Whitehead might use it, in the sense of the Greek Cosmos. Nature in the sense of Cosmos includes all living processes, divine, human, social, physiological, biological, and, finally, mechanical. Since, however, the physicists had not triumphed in his days, Hohenheim had not the faintest reason to give any preference to mechanics in his vision of nature. Mechanic processes were not, as most of us take for granted, the basic processes which explained all the others. They were—and by the way they practically are—only one sort of phenomenon in nature among a great deal of others which seemed to him—and seem to us again—much more significant and important for the understanding and interpretation of the creation in which we live than the laws of gravity.

This and many more ideas struggled for recognition. They lifted medicine to a new position in the realm of knowledge, taking over some chapters from theology, some others from philosophy, some even, in its social program, from the law.

To find a form for this new order of our world, this fundamentally new and biological *Weltanschauung*, was the real problem for Theophrast. He realized that, not one medicine only, but all our natural sciences, of living and dead substance (*ipsa philosophia et astronomia*) are without any solid and efficient foundation.

It was here that the real danger threatened. What he represented in the full vigor of youth and inspiration, by his record, his experience, his personality, his faith, his devotion to the poor and sick, all this had to crystallize now in a form that would survive persecution and death.

Modern scientists, writing about some special problems of one special field within the valley of the dead which is called nature today, are so unfair towards Paracelsus because they see no cause to relive his anxieties. How could he save the unity, the harmony, the lawful order of life, thought, art, that had happened to him, and that now was smashed to pieces by his defeat? He seems to start a hundred times to tell the whole truth. Each time he begins at another end; each time the whole, entrusted to him, him alone in the whole world, worries him and he tries to let us look, from his specific starting point, into the whole secret. Up to our days, the cosmos of man, spirit, nature of things, has not found a more comprehensive treatment. This seems a big order. However, when we look upon Descartes, Spinoza, Hobbes, Hume, Leibnitz—how many real, concrete processes of real human life, like birth, measles, friendship, despair, superstition, love, do they really tackle? Each one of them restricts himself to a very few. The selection of these very few first principles is significant for just this particular thinker. They follow each other, every one of them pointing to some particular principles.

Theophrast Paracelsus is different. Around the problem of life, every experience is amassed, and used as a means of approaching the central problem, and such is his reverence that the wealth of facts is preserved despite the drive to the center that animates all. Out of innumerable monographs emerges his general vision. We shall say something about his great concept of the unity of life at the end of our essay. Here we wish to mention one idea which never was appreciated in its fruitfulness and which like so many of the things he knew and practised, has a future. The theologians preached that the church was the body of Christ, with all the faithful as real members of this real body. Philosophers held that man was a microcosmos, reflecting the macrocosmos of the universe. Hohenheim speaks of the individual little man, and the great universal man. We all, he says, contain innumerable Egos because we pre-form, each of us, the innumerable potentialities that in the full life of mankind then are laid out in full size so to speak. Each potentiality comes to pass and is embodied in one individual or tribe or society. Still each individual is this same universality of all these forms, in a nutshell. In this way, Hohenheim reconciled the profound experience of the church, and the cosmic aspirations of philosophy in a truly human and truly biological and sociological conception. Not revelation, creation makes us members of mankind but so that any member is representing the whole kind, and still has a partial function too. Modern biology, with its cell-theory, says the same on another level. And the cellular theory again, is not very far from the tripartition of elementary processes that Paracelsus ascribes to all living substance, as we later shall see. In this section, we are concerned with the task of his last fourteen years. It was not that he had to write down "ideas, thoughts, theories". He had to transform into the poor form of manuscript his life, his work, his struggle, his mission so that a hostile and inimical world would be able to grasp what he stood for: A statesman can point to wars, conquests, treaties, laws, even when he is defeated. But how could this man point to the glory that surrounded him when he came to Basel?

So, in the most unpremeditated way, this doctor began to write on the sociology of medicine, on theology, on philosophy. Many of his theological writings still wait for publication. Our sciences may be compared to a tree, each branch having its proper day of origin, and serving a special purpose. Paracelsus, in his experience, seems to relive this tree of the sciences. For all of them are expressions of real departments of our existence. He did not overstep the limits of his calling when he expressed his actions, the principles of his travels, his motives. All this is inexpressible except in ethical, theological, philosophical and sociological reasoning. The humble servant of the art of healing was compelled to give birth to a full system of thought, the first system not recomposed from the headlines of the tables of contents in *Summas* and textbooks. This system was wrested from him, in the same way as Goethe meant it for his own writings, as "fragments of a great confession". The writings of Hohenheim are the first scientific writings produced as a life's fruit, wrested from him by necessity not by the external needs of a chair, an examination, a promotion, a functional usefulness within a school.

They were written on the highest level of self-defense where the Divine Inspiration had to defend herself against annihilation; and because her servant was disabled to operate as her legate, must be translated into his words as an author. We are observing the slow evolution of new literary forms, outwitting the departmentalization of the sciences as it then existed.

TWO PORTRAITS: 1526 AND 1540[25]

1526
by Holbein

25. *Ed.* https://www.astro.com/astrowiki/en/File: Holbein_Paracelsus.jpg.

ALTERIVS NONSIT QVI SVVS ESSE POTEST

EFIGIES AVREOLI THEOPHRASTI AB HOHEN
HEIM SVE ÆTATIS 47
OMNE DONVM PERFECTVM A DEO
INPERFECTVM A DIABOLO

1 5/H 40

1540
by Hirschvogel

The Rule of Twofold Beginning

Since the only level recognized by his contemporaries as scientific was lower than that of his new method, Paracelsus was defeated in Basel and had to live in mental exile for the rest of his life. How often may he have pondered about this verdict of fate. "Time is man's master", he says in his commentary to Hippocrates, "and plays with him as the cat with the mouse."

Still he took up the gauntlet, thrown against him by the world in Basel, and by doing so, the year 1527 became the axis of his life. Losing his office as a professor, he made his life the profession of the new office that he felt himself to hold. "A physician should be married to his art as a man is married to his wife, and he should love her with all his heart and mind for her own sake. They physician who is not married to his art with his soul is a quack, an adulterer, and an imposter." Weighty words for the man who is accused of just these three crimes. It shows what was on his mind.

In the 15 volumes of the admirable Sudhoff edition, itself a masterpiece of human devotion and wisdom, and finished a few years ago, one feels that each hour of the man's life found a place somehow or other, and was there condensed into spiritual life.[26] That a poet's works are the true children of his inner destiny, or should be, is a commonplace today. The logic of his works and of his life is one. And this truth is not limited to the artist. No dissociation of the living soul and the works of our hands is tenable. The secret of such a connection between life and doctrine was always felt in the case of Paracelsus. Thus, he became Doctor Faustus to some, a lunatic to others, the greatest alchemist, capable of making gold, the patron of Rosicrucians and secret societies. So far went the feeling that a colossal stone had fallen into the smooth pond of scholarship, in his person, that innumerable forgeries were written in his name. Mr. Preserved Smith, Mr. Dampier, and all the honest humanists who slander this great soul, all write, I have no doubt, in good faith. They glance over an unauthorized edition of his works, and they find mysterious and cabbalistic nonsense. Nobody ever despised this nonsense more than Theophrastus, and what is more, all the contemporaries that are played up against him in modern tradition, Erasmus, Bacon, etc., are far more infiltrated by the superstition of their age, exactly as we are by the superstitions of our own days.

It takes that indefinable newness and unexpectedness that we call genius to wage war against superstitions in a constructive way. So great was the gulf that separated him from his contemporaries, Luther

---

26. *Ed.* For details on the Karl Sudhoff (1853–1938) edition of the writings of Paracelsus, see online the University of Zurich's "Paracelsus Project".

and Erasmus, that only two or three men in each generation after his death took the trouble to seek contact with the real Theophrastus of Hohenheim. The year at Basel, then, created a cleavage, a bifurcation, in tradition. One half of Europe insisted on dealing with him after the fashion of Basel, calling him Paracelsus, ascribing to him every folly, later even inventing arbitrarily a certain Valentinianus whom he, Paracelsus, was said to have plagiarized. Thanks to Sudhoff, we know now that this manuscript was forged long after Theophrast's death. Anybody who admires men like Erasmus uncritically will belong to this half. Fortunately, Huizinga, the latest and wisest of the biographers of Erasmus of Rotterdam, does not share their prejudice at all.[27] The other half, of the people who care for Luther, for national politics, for religion, will overlook Theophrastus for his being so powerless in his own days. Luther, after all, made an immediate impression on everybody. Of Paracelsus, one might almost assert that his light was so far away from his incidental contemporaries that they did not see him at all. So, two halves of mankind can't place him in their picture of the world. They are ready to talk about natural science at a time and period when Humanism and religion retreat to the background anyhow. They do not see that a new form of thought must be lived first before it may be externalized into endowed institutions. And that is exactly what Theophrastus did: He lived that same life of immediate, encyclopedic, unprejudiced, experimental research on which modern society bases its existence.

It is unthinkable that this society could exist today without Paracelsus sufferings. Ninety years after his death, [Jan] van Helmont [d. 1644] basing his studies on the great edition of the collected works, came forward in his defence. The school of Paracelsists, for the next 150 years, fought its way into the medical schools. In 1750, they were in power. It is a stroke of irony, I at least feel, that Theophrastus finally should enter the schools by the means that he had despised in his lifetime and that are, however the only entrance ticket for the universities: by literary tradition. Van Helmont had not known him; he knew his books. He was able to quote him, volume and page. There it was in print, of old. And so it was good, and the advocate of immediate experience, who believed that professor and student could meet on

---

27. *Ed.* Johan Huizinga, *Erasmus of Rotterdam* (1924).

this basis only, had to be introduced to the houses of higher learning as a literary man, to be quoted from a book.

A few of his contemporaries sensed the truth; Pierre Ramus [d. 1572], for example, the great iconoclast who smashed Aristotle, would say: "So deeply did he penetrate into the deepest intestines of nature, with so incredible subtlety explored he the energies of metals and plants for the healing of every disease, even the desperate ones which mankind thought incurable, that with him as the first leader, medicine seems to have come into her own." And Melchior Adam [d. 1622], a humanist, well admitted that here was a man whose acumen seemed certainly, of a divine nature. Neither Adam nor Ramus, of course, were physicians.

If we transfer ourselves, by an effort of the imagination, into the years preceding the printing of his books in 1589, we shall realize that it is nothing short of a miracle that the force that had sustained Paracelsus himself in his titanic struggle through forty-eight years, after another forty-eight years prevailed once more. We therefore, may thrust our foot between the door and the threshold, thinking that here a glimpse into the functioning of reality is opened before us that transcends the average by its lucidity and importance. It is, for example, a great temptation to compare now the curve of life lived by Theophrastus of Hohenheim, with the rhythm displayed in the life of Erasmus von Rotterdam or of Martin Luther. Erasmus died in 1536, after having consulted Theophrast for his sickly constitution and re-paid him with one of his famous Latin letters. Luther died in 1546. Hohenheim died in 1541, he the Luther of Medicine as he was called. However, though it would serve our purpose of clearing the memory of our man from many misunderstandings, and of determining some-what better the part of a "Classic", a "revolutionary", and a "founder", we shall restrict this chapter to an interpretation of the rhythm in Hohenheim's own life. What a "founder" is—something apart from classic, pioneer and revolutionary—will, perhaps, become sufficiently clear even under this restriction.

Jesus founded the church, and in 325, this church triumphed for the first time. So, there is at least one historical process where a founder may be observed in his dealings with reality. Now, Jesus was no pioneer. He was fulfilling something. So absolute seemed the ful-fillment that he was never considered a precursor. In him the whole thing was achieved already. Nothing could be added nor taken away later. Leonardo da Vinci for the art of engineering, sometimes seems to me a founder. Everything we admire later in innumerable individuals—technicians, inventors mechanics, seems to live in Leonardo as in a cell, the seed of a big tree.

What about Theophrastus von Hohenheim, immortal under the nickname Paracelsus?

The life of our friend is divided into three distinct forms of existences: 33 years before Basel, unchallenged, unattacked, growing. One year at Basel, honored, placed, in charge of recognized social duties. Fourteen years, after this meeting with the established world of science and teaching, a target of slander, persecution, danger and illness.

The year at Basel, evidently is two-faced. When looked at from the beginning of this life, it is the fulfillment, an unbelievable opportunity to bring the right man into the right place. Looking back upon it from his deathbed, he might have dated back his early death, the tragic character of his life, to this same year. Theophrastus expired during this year, the charming, adventurous, generous, humorous and gay creature; Paracelsus began, the suffering witness of a great new truth, the responsible custodian of secrets which he never knew before to be imperiled. From the adolescent of childlike trust, the fighter, author, founder arose.

This is a thing described in the Bible as the acceptable year of the Lord; the highest times of men are whenever heaven and earth, world and inspiration, seem to meet. The formal appointment of a new professor coincided, in the year 1526, with an extreme case of inner readiness for an unexpected and absolutely new attitude toward science. For a short moment, external position and inner life seemed firmly balanced on all fours. He lived what he taught, and he taught what he lived. This messianic state of affairs never lasts. We cannot live what we teach, nor teach what we live, in the radical sense of Jesus or Parcelsus, since it is impossible to teach regularly for a life-time and because of the breadline, under those circumstances; and it is equally impossible to restrict yourself to a certain department of knowledge in your teaching if your real life shall be covered by your teaching. The simple fact of public or endowed schools prohibits the sale of inspiration day after day. A professional teacher or preacher is responsible for an institution, and not to his personal genius.

The acceptable year, therefore, always draws to an end. And when it ends, it will always entail tragedy, heaven ending, hell of despair opening, calvary and mortification around us.

A man who passed through the height of atonement between inner urge and outer requirement, to whom the harmony of ecstasy and duty, social and divine challenge became real, will utterly die to himself at the end of this period. By the very greatness of the event, he is put apart, separated from ordinary, natural men who always know how to distinguish between ideals and realities, and carefully insist on being called idealists by themselves and others and realists under the watching eyes of their parents and wives and children. All these divisions do not exist in the acceptable time. By a sacrament or a stigmatization—it really is both in a case like Hohenheim's—he is beyond the interests of the natural man. The pursuit of happiness is now meaningless to him. He looks at his own nature not less coldly than he does at any other partner in the game that is entrusted to him. He will use and exploit and outwit and overreach his own nature to make her the carrier of the message that is entrusted to him. He is in the know; he, then, cannot be called a pioneer, chopping wood in virgin territory, settling a country with his family, driven by an instinct of adventure westward, though a parallel undoubtedly exists. But rugged individualism and the pioneering spirit of property, are meaningless to characters who are dismissed from any prospect of personal happiness. When a man has gone through the absolute, when he is expelled from paradise, the power that makes him survive, is an objective. His own life is a tool now. For though he is nearly killed at first, by the catastrophe that always ends the acceptable year of the Lord, he is still there, very much to his own wonderment. And since he experienced the existence of the divine inspiration beyond any doubt just a moment before, suicide is out of the question. The same power that ruled harmony, is now proclaiming martial law; that is all he can see. Apparently, he is left as a witness of the higher life, as a herald of its promises and potentialities. Sealed with this indelible character, he is under one single obligation: What the world rebuked and refused to accept, has to be proved to be the acceptable gift of future life. Acceptable to God, not accepted by man, this dilemma contains a heroic challenge; for the tempter whispers, of course, smiling: neither man nor god is interested in your craziness. Under the spur of this inner temptation and the external disaster, the child of genius is turned into the fighting apostle. The disenchantment is complete. Few people will realize the degree of sobriety after the accepted time of grace led to the revolt of environment against inspiration. Whether the carrier of the inspiration survives himself, as in the case of Theophrastus Paracelsus, or his sister, as in the case of Nietzsche, where genius was protected by a goose, or in the great paradigma, where Peter, a liar, a weakling, a truculent fisherman, carried on the church, in every case, it is a perfectly rational, earthly, simple duty that is put upon the shoulders that survive. A precious gift being spilled, the drops of which are evaporating in a tragedy, it must be saved by all means. A virile discipline is required when loyal men establish a lawful order wherever an act of grace happens. The virility of the late Paracelsus complements the mirth of his youth; the beauty of his pictures as a youngster contrasts sublimely with his portrait, at 45, bald, pale, deep. (See our two portraits [in the preceding section]).

The tripartition of life, determined by the acceptable time in the midst of it, is common experience of all true humans. Only, it is strictly forbidden to voice this; for humanism does not admit tragedy in the middle of life. With great effort, it overturns the order of things; for example, we all talk as if the law came first, and grace and mercy afterwards. And it is true, when we take a crime, that this is the order of things: first the law over-awes the criminal, and he is sentenced to die. Only later may a governor grant a pardon. Here, evidently, the act of grace follows the act of the law. Since St. Paul discusses the Jewish Law and the Christian Grace at great length, it may be that this discussion also contributed to our confusion about the biographical sequence of nature, grace, and law. Paul himself is not responsible for this persecution of humanism.

In any life of normal health, grace comes first and the law follows. Any loving couple goes through the acceptable year first and out of this perfect happiness the special law of this marriage is derived and developed. Fluid flesh and blood precedes, and ossification follows. Jesus is free grace and his church is lawful order. Life is a process of crystallization. Free, revolutionary inspiration precedes; evolution, lawful development, is derived from the previous revolution and ecstacy.

Erasmus was a classic; Luther was a reformer. Theophrastus Paracelsus lived as the first citizen of a realm in which most of us feel at home by now. He lived a stranger in his time, without any hope of seeing success during his own life. However, he did not despair of his duty to transform his year of grace into the years of toil and lawful preservation. In this respect his life is far more christian than either that of Luther or of Erasmus. These two men taught and reformed christianity; Theophrastus added to it.

By a chain of marvelous concatenations and circumstances, he was brought up as a new type of man, moving in a new world, using new language, and living with his fellow men in a new fellowship. For an instant, he is placed so that his "newness and unexpectedness" become audible and visible to the world. Of course, no endowed institution will endure the contact with a creature that had never existed before. He is howled down from the chair, and the world does all it can to make sure that he will be down forever. He now faces despair, or compromise, or, thirdly, the slow road of waiving comfort, peace, and rest, and re-building, brick after brick, the palace of truth that before had appeared to him gratuitously. What seemed to come from above, as inspiration, now must be worked out piece, by piece, from the ground. He, despite this change in life's outward conditions, despite this complete change in manner, from eagle-winged flight to laborious plowing, keeps his faith; as Robert Browning, in the first poem that did justice to Paracelsus, said: "He is sure that God never dooms to waste the strength he deigns impart."[28] This way, grace is transformed into a new descendible law. We repay, by our faithful masonry, ploughing and building up from the ground, our load of gratitude for the inspiration, the abundance of inspiration that fills us in our best hours. Terms like grace, law, atonement, will, calling, perhaps get a new meaning when we re-read them in the light of such a life, full of revelation, full of grace and full of lawful response in the seemingly hopeless exertion of the man's last drop of strength. Why must books on the history of science or civilization be as dull as they are? Is it not because of the complete lack of ingenuity in our historians who never think that the life of Jesus is simply the law of life for all men, and what

28. *Ed.* Browning, Robert, "Paracelsus". In the *Poems of Robert Browning*. Vol. 1: 1826–1840, edited by John Woolford and Daniel Karlin, 127–135. London: Longman's, 1973.

is more, for all women. Therefore, they are without any scientific basis to work on. But genius has its everlasting, spiritual laws. As soon as we place grace where it belongs, in the center of life, as its inspiration, its directing force, life ceases to be arbitrary or accidental or casual and boring. And then we gain access to the writings of Paracelsus himself. He was aware of the true sequence of chapters in the book of life. In his Philosophia Sagax, he explains the strange fact that grace and free will are equally real. Superseding the vain controversies of later centuries between predestination and free will, he, in the small still voice of truth, says that grace and free will follow each other; grace establishes that law for which we sincerely can work with our free will. And the free gift of inspiration as he calls it, the Holy Ghost[,] is no contradiction to our experience of voluntary service later.

I only know of one modern writing about the same topic. In the "Meistersinger"—his wisest opera—Richard Wagner expresses this truth. The hero, Walter, is asking Hans Sachs: "How do I begin according to the rule?" Sachs answers: "You, yourself, set up the rule; and from then on, you follow it." We see, inspiration empowers man with sovereignty; still, man can prove that it was inspiration, no mere whim, by no other means than by submitting to this new truth himself.

One important conclusion can be deduced from this insight: any important thing in history is founded twice, once by a stroke of genius, a second time by the labors of duty. The United States were founded by the inspired Declaration of Independence, and a second time by the sober work of the Federalists. The Church was founded more than twice, but at least twice on Mount Tabor and at Pentecost. A college, a university, the Mormon State, a new science, anything worthwhile, is subjected to the law of a twofold beginning, one as a free gift from supernal inspiration, one as honest fruit of great fatigue and effort. The Bible, f. i. [?] Ex. 32, Numb. 11, offers cases of twofold beginning. And old Hegel, in his rationalistic way, stated it in these terms: *Aller Anfang muss zweimal angefangen werden.* The law of twofold beginning is the lesson taught by the fate, the experience and the results of Theophrastus Paracelsus. It explains and it connects many fragmentary dates in the history of human society.

Theophrast thus expresses our discovery: "When God withdraws his hand it is nothing short of taking the holy spirit from man and allowing him henceforth to act according to his own reason and his own pleasure. Where inspiration no longer dwells, there free will survives. For where the spirit listeth, all things must proceed according to him. But albeit that the creative spirit has withdrawn, still in such a man free will exists. And he loves this liberty from a free resolution of his will for the good and the choice of the good. And this man will not trespass the commands of divine inspiration from his own free will, now in his maturity. In the absence of genuine inspiration, then, two ways are open. One is the way of the damned. These people have the free will, too; though they have it for wickedness, for killing, stealing or betrayal. Whereas the righteous free will is his who passes the test in temptation, by his own free will, without the concurrence of divine inspiration."[29]

The Tripartition of the Good Life

The rule of twofold beginning is the rule of realizing, incarnating, embodying. The natural man, by the inspiration, is transformed into an operating force for purposes transcending him and his self. In Theophrast's own terms, the phase of inspiration and of free will in a man's life are similar to the passive and active form in grammar. During the inspiring vision, we are swayed off our feet, "and things proceed according to the inspiration". Whereas later, "man is allowed to act". "Man cannot create day," Theophrastus pithfully remarks, "nor can he create night, and he cannot create wisdom, but it must come to him from above" (*Labyrinthus Medicorum*). This passive reception which integrates us into an event in the history of the spirit, is one aspect and one phase only. For free will, in the midst of all the adversities of our environment answers the free gift by a rational obligation.

Since the prehension exerted by the inspiration, and the responding prehension exerted by man's matured will both operate on the natural man, the state previous to the passive and the active, the state of childhood, may best be defined as the middle voice in grammar. In this stage, man is already involved in a part of nature, by education, by environment. However, it is not yet decided what elements he will keep in common with his environment, which he will expel, and which he will add. Since no borderline is drawn between his nature

---

29. *Works*, XII, 421.

and the nature of his environment in this early stage, "medium" or "middle Voice" is the appropriate term for the first period of life.

Thus, the tripartition of the good life is elucidated. During childhood, or whenever we manage to be childlike in later periods, we are in that happy medium wherein we rely on the inspirations and obligations that animate the responsible members of society around us. We depend on parents, nurses, educators for food and shelter, physically and mentally. We ourselves can only hope for the best since we can do little. Hope is the deity of youth. We haven't yet brought down any personal roots into the life of the community.

Later, love befalls us. It takes possession of us in multiple forms. It commands us to take flight into a no man's land, that is to say into an adventure that was never tried before. Any passion that amounts to more than a physical tickling of the skin, forces us beyond conventions into a new constellation. A woman outside our clan, a task outside our traditions, a mission outside our country claim our dedication. The strength that is needed to face our environment and to break to it the great news that we are to be different from now on, we call inspiration. It is like the power of going uphill without panting. Everything seems so easy. It's a kind of flying, this honeymoon of first love. In fact, it is a real victory over the laws of gravity. Man is the up-hill animal of creation. We suddenly know exactly that the "middle voice", the innocent stage of the medium is at an end, that we are to be different, to become somebody different, for example a strange man's wife, a strange cause's champion; and we don't mind a bit. We don't fear the objections of our family and our friends. We smile at their warnings.

This absolute certainty that directs our steps is possible only because we are sure that the power behind us is bigger than our own weakness. We are precipitated, from above. Man being the animal that changes his environment, the phase of inspiration is that phase in which sufficient strength accrues to the individual so that he feels empowered to change the environment for the group. Inspiration does no more than that. It dislocates and places us. We cease to be a part of the environment, we are made the center of a new environment which, in our inspiration, is envisaged and anticipated by us.

As soon as this translocation is done, the honeymoon is over. We now have fallen into the new ground; we now are the seed that must be dissolved lest it bring no fruit. As the seed of a new form in society, we have acquired a new character. A woman, for example, just one of a thousand debutantes before she fell in love, now becomes this singular individual mother of these individual and singular children. This is indelible, irrevocable. She may divorce her husband; she never can divorce her children since a part of her own beauty and youth has gone into them. And Theophrast becomes Paracelsus.

In the process of contacting the new environment, of undergoing the painful birth throes of becoming a definite character, we very soon become aware of our own limitations. The infinite power, the radiant certainty leave us. The central choice whom to marry, where to settle, what to do with our lives, these great decisions appear with the force of manifest destiny. The petty daily decisions how much rent to pay, how to treat our wife's first cousin, how to arrange our courses, are open to reasonable doubt. The choice of our environment, we may say paradoxically, if it is to be successful, never must be felt to be our own arbitrary choice. The inner arrangement of this environment, however logical and simple it may seem, always will be more or less arbitrary and our own free choice. This third phase is a slow growth, in daily exchange and wrestling with the energies around us and against us. It is the slow growth of a man who knows that he means to stay, that he is in for good. A man, an institution, that originated in genuine inspiration, never will give up under pressure from outside. For they claim to have a destiny to fulfill, and in defending their mandate, they will demonstrate the nine lives of a cat. Their faith will prove invincible because it is a rooted faith, rooted in the previous experience of something bigger than one's own arbitrary and giddy choice. An institution, or a movement that deems its own faith to be but an answer to the prehension that determined it, is as a tree planted by the rivers of waters that bringeth forth its fruit in its season.

Surveying the sequence, we may grasp why children so long as they grow physically, as selves, can't be rooted into this world of realization. They still wait for the great affections of their lives.

And a professional enthusiast, too, is not a full man. He cultivates the phase of inspiration at the expense of that of incarnation. The everlasting idealist gives you the impression of a man who tries to prevent inspiration from ever coming true.

Finally, the "practical" man tries to live the third third of life all by itself instead of as the result of the two previous phases. His barbarism consists in taking his self-reliance not as an answer to the hopes of his youth, and to the love that characterized him, but as the first word by which he himself despite Paracelsus' warning, "can create day and can create night".

The fools of hope only, love only, or faith only, split the trident of our life energy. They pervert the fresh hopes of childhood into the sour milk of eternal moralizing, the great power of enthusiasm into nervous fits of excitement, and the tenacity of a responsive faith into the brutal energy of a "climber".

Man is apt to destroy the interplay between the leading three energies: hope, inspiration and free will. Most people think they have to worship only one of the three, and be ashamed of the two others.

Theophrastus Paracelsus discovered the tripartition of the good life and had the courage to be loyal to all three life-giving processes within him. For that reason he is no contemporary of the Middle Ages or of Modern Times. In fact, we easily are the first generation that may become his contemporary because we, for the first time, are faced with precisely his dilemma.

The preoccupation with Hohenheim is no luxury. It was natural that we founded a Paracelsus Society some years ago. All previous centuries were unable to approach the real and total man. They all picked out more or less external features. They were forced to admit certain contributions of Paracelsus to their organized work immediately; they never were able to admit the man wholly.

A short survey will show the gradual reception. In his own times, Humanism and Lutheranism dominated the scene. They were, like Socialism and Communism today, an evolutionary and revolutionary attack on the medieval cathedral of civilization. The Socialists of the 16th century, the Humanists, replaced the Christian saints by pagan heroes; the Communists, led by the violent Luther, left the visible church of the bones of saints and of stained glass completely

To his generation, then, Hohenheim seemed, at best, "The Luther of Medicine". Since Luther marked an exodus only from the stones and bones, and Hohenheim had grown up in the paradise of divine omnipresence in nature, the comparison was nonsense and resented as such by Paracelsus. He was not, by his antecedents, a protesting monk returning to the world after terrible struggles like Luther. He was a denizen of a living universe who claimed citizenship in the world of dead books and who saw his claim rejected. The world of witch-burning was not the world of our man; neither was the world of printed books.

The next generations turned from the stones and bones of the saints to the stones and bones of the real world. They took up anatomy, physics, astronomy. They ventured to touch directly the world of our senses. The sixteenth century slowly moved on the road to mechanics which were going to dominate all the following centuries. As to their method, nobody did so much for preparing it as Hohenheim, by the boldness of his wholehearted, reverent unprejudiced experiments. His concept of Chaos, that is to say, Gas, is one instance only of his exemplary influence. However, his method of strict observation was applied to dead matter only. The last four centuries will, on the whole, have to be called a period in which physics and mathematics dominated the thought of Western Man. Even God and the Law were proved by geometry. And the physical world, primarily was treated as a world of physics, of thermodynamics, electrons, or waves, or "bodies". All the sciences received orders from physics and mathematics, and are receiving them still today, directly or indirectly.

This arrangement means that we try to base life on death, the explanation of organism on the explanation of mechanism, and the processes in animated bodies on the laws of gravity valid for dead bodies. Modern science looks upon the universe as being a conglomeration of dead matter out of which by some unexplainable process, life may become developed in forms. In using terms like "body", or "energy" which are abstracted from living processes, physics was able to conceal the fact that it is decidedly the science of corpses, and of corpses only.

Now this certainly was not the world observed and disclosed by Paracelsus. Healing being his vocation, the integration of every process into a living universe was his great biological axiom. "Bodies" in the sense of physics, to Hohenheim were shells left over by life, and on their way to being recaptured by life. His method of reckless observation, then, applied to a much vaster universe than that of physics. As he once expressed it: his opponents seemed to see only one fourth of the real universe.

When his method, at least within the limits of the world of dead matter, was victorious, about 1750, it dawned on the world that the universe really was richer than geometry. Now it was not so much Hohenheim's method, but the size of his universe that, though very slowly, kindled the imagination. Organized science moved from physics to chemistry, from chemistry to physiology, from physiology to biology, from biology to phylogeny. But along this road, organized science still preferred to deal with the mechanic side of its subject matter. Death has always had the presumption in its favor during the last four hundred years. In other words: The complex universe, faithfully envisaged by Hohenheim in its totality, was recovered by science gradually, without accepting his axiom of a living universe.

This stubborn dealing with corpses and stones by natural science offers a striking parallel to the dealing with stones and bones in the medieval church. Both ages knelt in admiration over the relics of the past. Both evaded the issues of intense life in the present. The dogmatism of both ages put up a screen against reality. Today, Man seems to be unknown still.

Therefore a third stage in our relations to the living universe of Hohenheim seems to be reached. The whole range of his anticipations is perceived again. And we understand again what he meant when he treated the whole of the universe as the manifestation of a universal principle of life. A book-title like "A Living Universe" [1924] by L. P. Jacks may be rebuked by sceptics as accidental. These sceptics, however, should read the first publication in the series "Bios", Life, published by the leading English, German, Dutch, and American Biologists, in which the author, Mr. Adolf Meyer, adopted my definition of living beings[30] and our corroboration of Hohenheim's statement that the physicists only saw one fourth of the whole world. Professor Meyer explicitly relegates physics from its rank as the basic science to the background of a last and remote abstraction or ultimate generalization, a last frame for the ashes of the universe.

Biology, therefore, finally is facing the issue: Are we living in a living universe?

We are looking back today to the religious, the humanistic, and the naturalistic or mechanistic movement, all three, and we are compelled to live on beyond them all. And we find that long before,

---

30. My theory was developed in *Die Kräfte der Gemeinschaft* (=*Soziologie* I, 1925). It is accepted by Adolf Meyer in "Ideen und Ideale der biologischen Erkenntnis". *Bios*, bd. 1 (Leipzig 1934).

this man consciously lived the unity of the three elements: instinctive nature, divine inspiration, and reasonable free will which are put up alternatively by natural science, religion, and humanism.

Everywhere, Hohenheim shows his insight in the processes of incarnation. The ground covered by either theology, or natural science, or humanistic philosophy, does not interest him as such, but only as part of the whole process of life. The material of nature, the sublimity of revelation, the logic of pure reason—yes, of course, they are all there. And the only important question, to him, is their interplay. The fact that we shift from one state of aggregation into another, that life moves from naive hopes through supernal love into experienced faith and—in our children—back to hope again, that instinct, revelation, and reason are fundamental chapters of any course of life, is more important than the atomistic treatment of any one of them. Elsewhere, I was able to show the tremendous results for our conception of ethics and politics to be arrived at on the basis of this tripartition.[31] Here, we may show how Paracelsus' own biographical tripartition helps to elucidate an important point in his biology; possibly, this point will come to the foreground in modern research.

We found that Theophrast von Hohenheim ingenuously experienced the interplay of natural talents and instincts, inspired calling, and cold rational work. Now, he never tired to explain that life and any living substance was only possible as long as it was permeated by three elementary processes, mercury, salt and sulphur. It has long been understood that these names are confusing for us because we think of these three names as terms for "substances" while to Paracelsus they were elementary processes governing life lest it be incapable of corporification or embodiment at all. The conditions under which life can become manifest are the subject matter of his science, and of our study here. He says: "The three elementary processes are three forms or aspects of the one universal Will-Substance out of which everything was created. As long as these three are full of life they are in health. But when they become separated, disease will be the result. Where such separation begins there is the origin of disease and the beginning of death. To explain the qualities of the three it would be necessary to explain the qualities of the First Matter. But as the First

---

31. Rosenstock-Huessy, *The Multiformity of Man* (Norwich, VT, 1936).

Matter of the Universe was the "Let there be", the Living Word, who would dare to attempt to explain it?

Indeed we need no verbal explanation for the tripartition of energy reflected in Paracelsus' biography itself. Here, the facts of the man's own life furnish, not explanation, but illustration of his words. His life sponsors his doctrine. Such was the man's courage and wisdom and faith that he held one and the same truth for all creation and for himself. His biology and his biography are one.

By this translation of physical, intellectual, and spiritual processes into each other, Theophrast von Hohenheim really becomes the "Super-Celsus", the super-physician of our age. We are ill because the trident of instinct, revelation, and reason is broken in pieces. The divisions made by the churches and the sciences are untenable. In themselves, neither instinct, nor revelation, nor reason suffice, as regulating principles. Each has its time. To restore the process which leads life through all three, for its incarnation and integration, is the longing of our age.

In the identity of biology and biography, modern society faces the issue of its future. And in mustering all the masks of death, all the propaganda of physics, for nature; of creeds, for inspiration; of philosophies, for reason; we have some cause to despair. Our contemporaries have many a bone to pick with us. But where is life? Suddenly, we find that a man of 1527 A. D. went right at our problem. This experience has a surprising effect. It smashes the iron prejudice that, after all, a man four centuries old never may be our contemporary. Paracelsus is our contemporary much more than most of the men who must prove this quality by their birth certificate. And for that very reason so many of his modern critics simply are behind Hohenheim's time. They are obsolete compared with him.

This, then, is the last conclusion of our study. The tripartition of life has an effect on its duration. For whenever it is achieved in a man, in whomsoever these three life-giving processes had their full sway, life is sublimated into a form that remains of vital importance beyond the lapse of time. The carrier of such tripartite life is our contemporary forever.

Scientific Bad Humor (Bibliographic Confessions)

The bibliography of Theophrastus is in itself the greatest adventure in books. Its peculiar character was revealed by the master of all who know in the field of history of medicine, Karl Sudhoff, first in his two volumes, Paracelsus *Bibliographie*, 1894 ff., later in his monumental edition, in fifteen volumes, of the writings of our hero, with the exception of most of the theological manuscripts. His introductions to each volume are goldmines of information.

Miss [Anna M.] Stoddart, in the year of her death, published a charming book in English. This publication of 1911 [*The Life of Paracelsus* (London, 1915)] is out of print now. It is the only fair representation in English of the real Paracelsus. For example, she is the only writer that mentions how Lord Lister was anticipated by Paracelsus: "Keep a wound clean and open, and it will heal."

Browning's poem [*Paracelsus*, 1935] will always remain a great document though he read in Paracelsus a nineteenth century Byronism quite abhorrent to this humble servant of the poor and ill. [Erwin G.] Kolbenheyer's novel [*Das dritte reich des Paracelsus* (Munich, 1934)] is groping after something important. [Friedrich] Gundolf [*Paracelsus* (Berlin, 1927)] remains a purely academic performance; and probably was not intended to be more.

Two Austrian scientists, Franz Strunz [*Paracelsus; eine studie* (Leipzig, 1924)] and Franz Hartmann [*Life of Philippus Theophrastus Bombast of Hohenheim* (London, 1887)] contributed considerably to the understanding of the physician and the scientist. A great piece of literature is another physician's study, Victor von Weizsacker, Hippocrates and Paracelsus [*Bilden und Helfen*, 1926].

A short abstract of Fritz Medicus, "The scientific significance of Paracelsus", was translated in the *Bulletin of the Institute of the History of Medicine* (1936) IV, 353–366.

A "Paracelsus Society" was founded five years ago, in Munich. In the general histories of Science, the only serious effort was made by Emanuel Rádl, in his History of Biology [*History of Biological Theories* (London, 1930)], to expiate the ludicrous performances that dishonor the scientific tradition of our times.

Since it is an important part of reality, this centennial bad humor against the 'Faust'-type must be illustrated by some examples that, at the same time, will help to explain how the scientific process is nothing merely mental or abstract but the vital process of man and mankind itself, concerning the whole of man's personality and character, vitality and faith.

As the standard bearer of humanism against Paracelsus we may mention Andreas Libavius, in his Anthology on Alchemia of 1597. Here, all the great achievements of Hohenheim are turned against him as either diabolical or lunatic as follows:

1. Paracelsus did not respect the departmental spirit: "He united chemistry and medicine and," Libavius exclaimed; "hereby reversed all the sciences."

2. The great statement of the Basel program is recriminated—after 70 years!—again and again. Paracelsus had made three simple points:

   a. Experience shall guide me.

   b. I am myself the author of the texts on which I am going to lecture. (*quorum ipse sum auctor*)

   c. I am lecturing in German.

About "a." Libavius exclaims: "May he remain by himself. Authority means more than experience." Accordingly, his own book is made up from a list of some fifty authors of all times and places.

As for "b.", he tries to be very witty: "*Non quidem repudiavi si quas formulas apud Paracelsum inveni quarum fors ipse auctor non est.*" I have not repudiated formulas that I found in Paracelsus the author of which he perhaps isn't.

About "c.", Libavius moans: "If they would not, in their lunacy, prostitute (one of the pet phrases of the set mind) sacred medicine by German versions, medicine would stand in better authority." According to this typical eclecticist, "Paracelsus is a delirium, deserves no authority, his writings are impious against God, pestilential, filled with horrible lies of world and God, and Paracelsus is guilty of blasphemy and no vote can absolve him. The filth of Paracelsists . . . but already too much has been said about this Cloaca." *This is pure poison.*

Nevertheless, it is important to note, that the reader has before him, in our quotations, the whole substantial material which Libavius was capable to produce against Theophrast.

Later writers were equally venomous. Everything that the human mind may invent, was invented against Paracelsus. Any group in society seems to need one permanent scapegoat who has no rights whatever.

"He seems to have written his books in a state of intoxication". I. G. Zimmermann. "As a boy, he was maimed, and hence, was a castrate. He was epileptic. He was insane." K. G. Neumann.

The man's worst enemies, it is true, were his henchmen who merely sought refuge behind his powerful name. Scores of forged manuscripts were put forth, advertising the same nonsense in alchemy, astrology, sorcery, mysticism that Paracelsus fought tooth and nail during his life. Undefended by the profession as he was, unprinted too, he fell the victim of the well-known technique to smother a man under false laurels. Since this process was told by Sudhoff, it may suffice to expose the most recent example. In 1933, the Masonic Supply Company of New York published a book on "Philippus Theophrastus known by the name of Paracelsus". As the title shows, the author knows his stuff well. However, his mystic public expected miracles. And so on page 108 this beautiful derailment happened:

The reader will remember that, in his Basel days, Theophrast was annoyed by Pasquins one of which was a letter sent up from Hades by Galenus complaining of the revolutionary disturber of his peace.[32] The intermezzo was known to us because Paracelsus made fun of it in the preface of his Paragranum: "O the soul of poor Galen! If he had remained faithful to experience (Theophrast's guiding star) his remnants would not now be buried in the abyss of hell whence he wrote me a letter. Such is the fate of all quacks." Put on the track by this allusion, the indefatigable Sudhoff discovered the Pasquin, a Latin poem in clumsy rhythm, and printed it many decades ago.

The joke of Theophrast is taken up by the modern Theosophers as a revelation: And we read: "It appears from this sentence (in the Paragranum) that phenomena of modern Spiritualism (entering en rapport with a certain mind, writing or speaking in the spirit of a deceased person) are not a new revelation, but were known and explained three (4?) hundred years ago."

The enemies of Hohenheim in the 16th century were great forgers too. For example, they invented a complete author, Valentinianus,

32. *Ed.* "Pasquins" are satirical verses, lampoons, attacks by way of ridicule.

who was said to have been plagiarized by Paracelsus for everything that was of any value in the latter's writings. Though this forgery was proved long ago, modern books on the history of science still go on quoting this antiparacelsean invention as a genuine source.

Today, it is not so simple to omit Hohenheim in a textbook. On the other hand, he fits so badly into the list of humanistic and mechanistic Saints canonized in the 18th and 19th century, he contradicts too many glories especially that of Bacon of Verulam. French and English political history, for reasons that I discussed elsewhere at greater length, are unwilling to recognize the chronology of the German Reformation, and date the Renaissance correspondingly too late. Paracelsus does not fit in the scheme of the enlightenment. Vesalius and Leonardo are enthroned instead.

The modern historian of science, mostly unconsciously, is laboring under these political and religious prejudices. Certain points are repeated in our handbooks again and again though they were refuted by Sudhoff, Rádl, Darmstaedter, Strunz, Hartmann, Miss Stoddart, Richard Koch, myself, long ago.

First, of course, that he called himself Paracelsus. Second that he wrote a bombastic style whereas he created the first scientific German prose, in a pithyful and simple manner, an abomination, it is true, for the Latinists.

He did not burn the Arabian medical books of Avicenna in his classroom. He originally was a good fellow, helpful and polite. After the catastrophe in Basel, he, for years, was in a bitter mood, and tried to explain his own position, as distinct from the Galenian, in prophetic and violent language, which, however, has to be measured by the language of a Luther, an Aretin [Aretino], a [Ulrich von] Hutten and which is the simple truth in every material assertion. He is attacked today because he warned people against imprudent operations. Here a modern issue, between conservative and bold surgeonry is simply carried over into the past. Why should Paraclesus have delayed the progress of science because he stressed—in the year 1536—the healing powers of man's own nature?

Some of his many discoveries in chemistry are: determination of the amount of iron in water in gallic acid. He was the first to claim zinc as a particular metal, determined *alaun*, used mercury, zinc, laudanum and lead, was the first to produce psamech paracelsi (tartar), arsenic acid. Sulphate Potash was first prescribed by him. He advised the vapor-bath for distillation. Ether was used by him as a narcotic before others, and he described its effects. He preceded the Italian Girolamo Fracastoro in the scientific treatment of the Morbus Gallicus. When Hohenheim had finished his work, full of medical prescriptions that prevailed for many centuries, especially in the use of mercury, its publication was prohibited[33]; instead Fracastoro published his poem that invented the euphemistic name Syphilis and won a reputation.

Hohenheim knew, in sharpest contrast to all his contemporaries, the truth to which physiology returned at the end of the nineteenth century: that "in the human being, there is present an invisible pharmacy and an invisible physician who produces, prescribes, dispenses and administers suitable remedies as occasion demands. Had not God created them, then notwithstanding all the efforts of our physicians, nor a single creature of the earth would remain alive". Everybody knows that this is a great truth, so much so that Bernard Jaffé in his *Outposts of Science*, 1935, when speaking of a modern explorer of the glands, of Abel, sums up Abel's position in the one sentence: "Abel felt that the words of Paracelsus were true (p. 162)."[34] Paracelsus perceived (to continue our list of his achievements) that air was a mixture and that gases—what he termed "Chaos"—were something far more general than air.

He rebuked astrology and said that the stars had no influence on life on earth. He tried to express the process of life in biochemical terms—exactly as our biologists today. His tripartition of the archeus into three material processes, all balancing each other, is neither refuted nor surpassed in its epistemological depth and its divination of the laws and categories of human understanding.

He conceived of the calcoli as one great process of dross throughout the whole system.

He investigated the use of magnetism for cures.

He wrote that Biblical medicine was very poor because Moses had other things far more at heart.

He was the humblest of the humble when learning from the common man was concerned, and his charity toward the sick and poor,

---

33. Sudhoff, *Werke* VII, 23.

34. *Ed.* John Jacob Abel (1852–1938), an American, is known as the Father of Pharmacology.

his valiant fight against graft in hospitals and pharmacies is on record. Many other merits are ignored even by his scientific admirers because they themselves are specialists in one field; for example his distinction between hereditary and un-hereditary talents goes unheeded to this day.[35]

Now we shall observe how this pure and devoted and illuminated life is pinpricked by the moderns.

Benjamin Ginzburg, *The Adventure of Science*, 1930, mentions Paracelsus only once; for what purpose? to say someone else refuted his theory of magnetism.

A. Wolf, *A History of Science, Technology and Philosophy in the 16th and 17th Century*, London 1935, takes no notice of Sudhoff's standard edition 1919–1933, nor of Huser's, but quotes the spurious of 1658. He gives, on page 445, a list of the famous physicians that, naturally, includes Hohenheim. Then, however, he goes on saying that Hohenheim's life was mentioned before and that therefore the lives of some other doctors are told by him now. Unfortunately, he is mistaken, and the life of Paracelsus is not told anywhere else in the book. The humanistic wit Fracastoro, probably because he wrote in Latin verse, gets full treatment. The much younger Ambroise Paré (1510–90) is singled out by Wolf to hit Paracelsus, in these terms: "A son of the people and no scholar, Paré was ever ready to learn, even from old house wives, and in this way came to adopt such remedies for instance ... as raw onions. The modesty of this great doctor and surgeon forms a pleasing contrast with the bombastic attitude of Paracelsus."

So, Paracelsus, a man of the people plus a scholar, is humiliated by a man who not only followed in his footsteps, but made a brilliant career where the pioneer was persecuted. The Plato of biology must be measured by the Calvin Coolidge of medicine; for Pare' was simply a doctor trying to cure his cases. Hohenheim tried to express a whole new order of the world, in the new light of nature.

W. C. Dampier-Wetham, *Cambridge Readings in the Literature of Science* 1924, p. 74, prints from a most discreditable, theosophic source and translation, instead [of] from Sudhoff, a doubtful text of Paracelsus, and adds: "His writings well illustrate the characteristic confused treatment of scientific problems by the later medieval mind, before the Renaissance cleared the air." Here, everything is turned

---

35. *de artium et facultatum inventione, Works* XIV, 253.

topsy-turvy. The man who "cleared the air", literally and metaphorically, was Paracelsus. Paracelsus attacked the Renaissance and Humanistic medicine on grounds we are just now reclaiming for our science of the living. For the Renaissance was given to a stolid Galenianism, in reverence to the classical texts.

As we saw, A. Wolf omits Paracelsus' biography. Still, he has to mention him occasionally by inference. How does he treat him there?

344: "In the meantime Lower had also adopted the Paracelsean idea of the composite character of air." Nowhere has he stated before, under the name of Paracelsus, that this idea was conceived by him. Most illuminating is Wolf's treatment of the first great Paracelsist, van Helmont. After having mentioned Paracelsus as a mere name on p. 325, he gives Helmont's life on page 326: "It was in this way (of medical service to the poor) that he came under the influence of the medical chemistry of Paracelsus whom he greatly surpassed. Van Helmont's greatest service to chemistry consisted in having been the first to show scientifically the material character of gases and their variety, the term 'gas' was actually introduced by him (he derived it from the Greek chaos, an expression which Paracelsus had applied to air)." One stands gasping. So wonderfully is truth and slander mixed.

1. Paracelsus did not call the air by the new name chaos; but used chaos because he understood the composite character of air, and needed a more comprehensive term, i.e. the later "gas".

2. As the only instance in which Helmont "greatly surpassed" his master, and in which he showed himself to be of great service to chemistry, we are told of a discovery which Helmont learned from Paracelsus, and therefore, was able, in the comfort of his station, to develop. It is, by the way, the only thing always credited to Helmont, as against the score of important innovations made by Paracelsus.

3. Van Helmont not only came under the influence of Paracelsean chemistry, but the man's whole personality, and always freely acknowledged this discipleship, a fact though not denied yet carefully omitted by the phrasing of Wolf.

4. The clause: "Whom he greatly surpassed" is set before the reader without any further reference to the own merits of the belittled

at any other place of a work which promises a history of Science and Philosophy of the 16th and 17th century.

The hatred of these rationalists is indomitable. The architecture of another paragraph is a masterpiece in this respect: "Paracelsus, it is true, denounced the association of astrology with medicine and proclaimed that the stars control nothing in us. But he only substituted for it his own equally delusive fancy when he added that the archeus not the stars, control's man's destiny." I hardly believed my eyes in reading this.

According to Wolf, the "Mneme" of Semon, the "Gene", of Morgan, the law of Mendel, the principle of selection of Darwin are "equally delusive fancies" as astrology. Any working hypothesis for biology inside the organism itself is placed on the same level with astrology.

Preserved Smith, in his *Age of the Reformation*, surpasses even Wolf. In his bibliography, he quotes seven works on Leonardo, five on Copernicus, not one on Paracelsus. He ignores Sudhoff. No wonder that he reports from more hearsay in his text: "The greatest name in the first half of the century was that of Theophrastus Paracelsus, as arrant a quack as ever lived, but one who did something to break up the stronghold of tradition. He worked out his system a priori from a fantastic postulate of the parallelism between man and the universe, the microcosm and the macrocosm. He held that the Bible gave valuable prescriptions, as in the treatment of wounds by oil and wine."

The Microcosm-Macrocosm parallel is not Theophrast's brand at all. He corrected it as we have shown in the text. And he did this, after forty years of restless toil and ever renewed experience, at the end of his life. He stated clearly that all Hebrew medicine was unreliable because Moses was interested in theology, not in physics (Liber Azoth, chapter on human bread),[36] and because Israel did not take real interest in this world. And was, in fact, the first to teach Asepsis.

But what about the biographer of Erasmus, Mr. Preserved Smith, who knows so well that all contemporaries of Hohenheim believed in the inspired letter of Holy Writ? Hohenheim was one of the very first to criticise the Hebrew tradition. Why then, sting a man who is the

36. *Ed.* "Liber Azoth" refers to writing by Paracelsus but the reference is vague.

least credulous of all, with a reproach that applies to every orthodox Christian down to 1859?

Except for a slovenly remark a hundred pages later, the quotation given here, is all the information about Paracelsus, in a volume of 850 pages on the Age of the Reformation.

The only really remarkable thing is that these detractors all are compelled to call their despised victim "the greatest name" or something similar. They are nothing but the prolonged arms and thoughts of Paracelsus' humanistic contemporaries among the physicians. Mr. Preserved Smith is not even in his biography of Erasmus mentioning the fact, that Erasmus himself consulted Hohenheim.

Scholarship is not an achievement of "the empty intellect" (Faraday) but of living, fighting, loving and hating persons. The scientists are divided into the two groups of those who admit and those who repress this fact. Real lives try to live the source—life of the heart; to put their hearts into something important! They risk to be destroyed through the persecutions of the other group that boasts of being pure 'mind'. For, the mere mind, by ignoring or fearing passion, is unable to integrate mental and passionate processes into a whole, and hence becomes unable to master the passions. Of course, the passions are not annihilated by ignoring them; only it is true that they are perverted by being denied. When the constructive passions are declared not to exist or to be bad taste by the pride of reason, they will turn into hatred, and lead to outbreaks of hatred within the realm of science itself. The Rationalist's bad humor is a reality, and an important reality in the process of science. Consequently, a world-heart such as Theoprastus Paracelsus who challenged all the world to share his whole and primary life raised against him all the powers of derivative and divided life. Their defense-mechanism is at work against this great soul for four hundred years. And so we may learn what is meant by the powers of hell. They are raised when the powers that be are not conscious any longer of the fact that they are derivatives from the primary powers of the heart.

Part III. The Common Denominator for Classic and Founder

EXTERNAL DIVERSITY

Obviously, classics and founders of science fulfill a different function in the growth of a science. Their relations to society seem to be almost opposite. Faraday met with praise and appreciation; Paracelsus was persecuted and nearly destroyed. The science offered by Faraday was eagerly expected and greatly admired; the science envisualized by Paracelsus was feared and declared impossible. The increase in Knowledge through the work of both men cannot be measured by any objective yardstick. But the data given by us about Paracelsus show that the change in knowledge made by him in a short life would have meant a greater revolution in science than even Faraday's discoveries, had they been listened to, received and digested by his age. As it was, no such possible progress was made.

This obvious contrast in the situation of natural science within society in 1526 and in 1820 is clearly a determining factor in the making of a "classic" or a "founder". For its clear definition, it may be helpful to admit at the start that personally a classic may meet with all sorts of difficulties and hardships, and a founder with an abundance of social advantages; and still, they will depend on the objective phase of social evolution. Therefore, we see, between our two heroes, the more tortuous and difficult career on Faraday's side. He was poor and unknown and uneducated and Davy's valet. Similarly, the great "classical" contemporaries of Theophrast von Hohenheim, Michelangelo and Erasmus, had a harder youth than Paracelsus. The latter was the son of an academic physician of good standing and social reputation.

This admission does not diminish the importance of our statement that a classic's achievements meet with universal appraisal [approval?]. For this only means that one cannot become a classic without an atmosphere and a public which expects and welcomes the advancement of this special field. Richard Wagner was hated during his life, but music had her heyday in his time. He did not live in Plato's republic where music was forbidden and execrated. Wagner lived after Bach, Mozart and Beethoven. He had to fight for "his" kind of music, not for "music". With regard to the classic Faraday, we find that physics, for two centuries had made a gradually deepening impression on the European mind. Therefore, with Faraday, physics themselves [itself], finally, gained social and cultural recognition, even from the common man, as a universal blessing and an asset to humanity. Faraday's aim and field were welcome.

The blunders of the historians of natural science concerning Paracelsus are easily explained when we consider their negligence about the timing of science. They treat any one scientist as an individual, and try to define his character as an atom in the universe. Then, it is true, the fervor, the anxiety, the sacrifice, the dangers of a genius like Paracelsus become wholly superfluous. In an alphabetical index or in the Dictionary of Biography, every individual seems safe; his individual contributions are listed; all scientists, poets, etc., seem to work more or less on the same level. The problem of Paracelsus was not at all "how to make his contribution". It was, rather, to enable scientists to make contributions into a new reservoir, a new system, a field hitherto not defined at all.

The founder may be liked personally and belong to society. Paracelsus was the equal of his colleagues from the outset. Only, his plans seemed absurd, his aims ridiculous; his vision seemed madness; his new scientific "Monarchy" of experience in biology sounded as blasphemous to humanism as a science of the living soul sounds humbug to modern scientists. Not the man but his intentions were undesirable. Any founder is a failure in the eyes of most of his contemporaries as compared to his immediately successful competitors. May be that he lives to see some public recognition at the end of his days. This depends on the accidents of his physical vigor; mostly congratulations which are tendered to an octogenarian are purely accidental when related to this man's struggle at 35. During the periods of his greatest effort, this approval is withheld, because the eyes of men never can see without love or hope or promise pre-organizing them. Romulus founding Rome was not a success in the eyes of his contemporaries. Yet, these same contemporaries, probably including his brother Remus, thought of a senator of the neighboring towns of Tarquinii or Alba Longa as completely successful. And even the founding of a new place by Romulus was routine work when we compare it to the enterprise of Paracelsus who enchained the stream of new sciences, in their own right, and in a ceaseless procession out of the womb of time, in an environment spellbound by classical books. His grain of seed was far more inconspicuous and unrelated to anything previously practiced.

And yet, though every external element differs in the destiny of classics and founders, these differences between hissing and applause, invisibility and luster, fade away as soon as we analyze the real merits of classic and founder. Both, classic and founder prove the same laws for the mind in action. In temper, habits, speech, fate, Faraday and Paracelsus certainly have nothing in common. All the more astounding is the identity of their heart and soul of which we have to speak now.

INTERNAL IDENTITY

The first law proved by their lives, is the fact that natural science presupposes one common education that comprehends the people and the scientists as well. Scientists must find themselves integrated into society before they can set out for their special functions; otherwise society will not stand up for science.

And when this common faith does not exist, the scientist himself has to step back patiently and create the soil of a new public faith for his plantation of a new science. In the days of a classic, this soil of a common faith exists, and spares the scientist the dividing of his energies. The impotence of modern scientists to understand Paracelsus originates from their ignorance of this law. They did not see that a founder has to do both; create a new soil and plant a new tree, witness a new faith—for the general·public—and at the same time sow the seeds of a new knowledge. In this twofold role, his creation of the new faith is not understood by his alleged scientific colleagues: they deplore his walking with the sinners, the laymen, the uneducated classes. His scientific efforts are absolutely inaccessible to the common man with whom he shares his new faith.

Reason can't build a body of science or a republic of scholars, before the hearts of men are trained for the corresponding equilibrium between future, past and present. In order to place the house of science between the future and the past, Paracelsus had to raise inaudit [unheard of] expectations of a future unbelieved by theologians and humanists. The theologians talked about the end of the world or the Anti-christ. The Humanists hoped for a second Antiquity. The faith of

Paracelsus was the emancipated faith of an adult, who, on the basis of the previous Christian revelation, now experienced the day of revelation in nature as a further chapter in the inspiration of mankind.

The very existence of science, and all the more its steady progress and perpetual regeneration, presuppose an efficient social education. For education connects men of different interests and aims in a common faith as to the direction of society. The classic, Michael Faraday, and his European public in the 19th century, were sufficiently steeped in a common faith, Faraday drawing infinite resources from the Christian training of his heart, and his public having that faith in science which had been created by Paracelsus, the Paracelsists, and the later physicists. And because of this mutual permeation of his own and his public's faith, Faraday thrived and grew like a tree.

Theophrastus von Hohenheim was made into Paracelsus; the vigorous genius of the picture by Holbein was turned into the man of suffering shown by our second portrait in 1540, not more than fourteen years later, less because his contemporaries were not scientifically minded but because they were foul-hearted. And he met with foul play for another four centuries because few people admit the relation between faith and science, between future and presence.

Natural science cannot thrive in the void. It owes its opportunities to the faith of an integrated society. The tragedy of Paracelsus was not in vain if it destroyed the misunderstanding in our own minds that education can be based on science. This is the most popular assumption of our age. Nevertheless, it is not true. It is, of course, a truism that the mind may be and should be well trained by scientific methods. Only, we have to be in agreement on the meaning of "well" trained, of truth, of solidity, long before we are able to use the specific methods of biology or physics for achieving our aims with our children in this direction. Why do we love truth, the training of the mind, independent thinking? Perhaps a society may prefer patriotic lying, hazy enthusiasm to our campaigns for clear thinking. A predilection for mendacity is frankly avowed by the newest social creeds. When they enthrone a profitable mendacity against a science for its own sake, they must be met not by better science but by a deeper faith. Science is based on faith, on a very specific faith, perhaps, and the different sciences all anticipate different aspects of mankind's destiny. However, all occidental science is more than curiosity; it is carrying out a sacred obligation, it is fulfilling the prophecies of old. And though we may need a new branch of the sciences today, the now already old natural sciences and these new ones, both, must be based on the common faith of mankind in its destiny. Natural science has to go back to its founders in order to restore its own accounts of its activities to their original meaning. In my quality as a scholar, I naturally am tempted to go ahead recklessly with my reasoning power and to laugh off any suggestion that scientific progress presupposes a solidarity of heart and soul between the know-nothings on the one side and the know-much on the other. But then I am reminded of the vicissitudes in the march of science; where, as in Darwinism today, a central idea like evolution may be breaking down after a triumph of fifty years. And it becomes clear that no society can be based on any content which is shifting so rapidly as scientific theories. Any scientist tries to carry over his opinions into the field of social reality as reck-lessly as possible. But that is just our temptation; we have to resist this temptation or we are digging ourselves the grave of science.

Science, as a body of knowledge and as a strategical campaign of the human mind, is a social achievement and as such, science is based on social education. The people must be integrated and come to feel again and again that they all are identical in heart and soul lest they withdraw the Magna Charter of scientific doubt bestowed on Reason by the great-heartedness of society. The heart of the scientist must remain identifiable with the heart of mankind. As long as this fact is respected, the scientific mind may set out for his adventure. The loss of this identity kills the life of science and of society.

This solidarity of the scientist with the very heart of humanity was lived by both Paracelsus and Faraday.

In the case of the radical rebel against textbook humanism, this solidarity with mankind's destiny had to take precedence before sci-entific success. Paracelsus clearly knew that he defended a new faith from which new sciences would spring like locusts. He had lived this faith in the marvelous decades of his unique youth, those years of pure experience in three different worlds. And so he insisted on the proper hierarchy between heart and mind. Otherwise, he might have been absolutely successful with his colleagues; he might have made them listen to his discoveries simply by linking his new discover-ies, as they did, to old Latin texts in one or the other tricky way of interpretation. There were certain techniques which allowed a doctor to innovate by compromise. Only, by remaining within the world of humanistic shadows, Paracelsus would have sacrificed his truth, his vision, the very sap of his existence, to social success. In a dilemma such as the dilemma of Paracelsus, between world-heart and world-mind, the social failure of the man becomes the test for the success of his new foundation. The failure of the founder is the condition of his work's success. The failure of the founder and the success of the classic are one and the same act, performed at different stages of evolution. They are different avatars of the soul of science. Our two heroes are outstanding because both reveal the whole man, heart and brain, soul and mind. Both link the profession which they create or represent to the universal tree of humanity. Both embody the right hierarchy of values. Surprising as it may seem, the classic gave evidence of the same law borne out by the founder that the body of scientific doubt must be rooted in a living and undebated social faith.

| The classic is the fruit of this faith and embodies his science. | The founder embodies this faith and is the seed of his science. |

The combination is different; however, it is a combination of the same elements. When we apply the rule to our two cases, we may express it more concretely in these terms: Faraday is Faraday because he embodies his science; his faith though of first-rate importance as a condition is not his principle of individuation. Paracelsus is Paracelsus because he embodies our modern faith. His scientific genius in medicine, biology, sociology, chemistry, education, though an important condition is not his principle of individuation. Richard Koch, and Victor von Weizsaecker, turned to Paracelsus when they wished to gain clarity about their own faith as modern physicians. Abel turned to him as a biologist. Goethe and Browning, instinctively, turned to him when they tried to express their own human faith. The "Faust" is, after all, the sublimation of the popular legend of Paracelsus. Finally, Oswald Spengler, in his Downfall of the West, called the whole millennium 'Faustean'; in doing so, he exalted unknowingly Paracelsus into the embodiment of Western Man in general.

And now, we are in a position to understand why both, the classic and the founder are not explained, in their role among us, by the yardstick of external success or failure. While it is true that one meets success and the other meets failure, their common denominator consists in their readiness to accept the one or the other as mere by-products of life. Martyrdom and success, both, are coveted by certain types of man. Faraday and Paracelsus are too vital for being 'typical'. They are persons in the making. A person is beyond the typical. Martyrdom has no merit per se, and success has no merit per se, for a living person. The founder may get very near the stake where witches are burnt, the classic very near the throne of coronation—both are little concerned with these precipitations of their process of living into external moulds. Fortunately, we have striking biographical material from both men from which it can be proved that a victimized founder must be carefully kept apart from a martyr-volunteer and a classic carefully distinguished from the merely successful man. During their whole life, these two made a real effort to evade any misunderstanding in this respect. A fanaticist would proudly say: 'I am a martyr'. Paracelsus, instead, wrote a beautiful chapter against voluntary martyrdom, in his booklet on invisible illnesses (edited by R. Koch and myself in 1923). In this chapter, he makes fun of the ranters who triumphantly run to the stake of martyrdom as though it was a bonfire. We ought to remember his being hounded like an outcast when he wrote the lines; then, they gain momentum. And Faraday did not say: 'I shall be successful'. He declined decorations, two Presidencies, a knight-hood, and a sure fortune of 750,000 Dollars, all offered to him. "I must remain old Michael Faraday to the end", he would say. Both men were tempted as we see, to skid off the narrow ridge of freedom into the valley of their pseudo-type; the boastful failure or the boastful success. They were more than martyr and more than conqueror. Any man is tempted in his career to establish himself in his external garb firmly. The founder undergoes the real temptation of posing as a victim. And to pose as a fake-conqueror is the temptation of the classic. In order to keep alive, a man has to discriminate against the external marks of his mission; only so may he uphold its inherent truth. Paracelsus, it is true, became a martyr of his living faith at least in the same deep sense as his contemporary Thomas Morus became a martyr of the Church. However, though this turned out to be his historic experience, he always remained superior to this role valiantly. When we read his great—and neglected—Philosophia, one of the most personal words of Shakespeare keeps running through our mind:

> "He's truly valiant that can wisely suffer
> The worst that man can breathe; and make his wrongs
> His outsides,— to wear them like his raiment, carelessly;
> And ne'er prefer his injuries to his heart,
> To bring it into danger."[37]

Hohenheim became Paracelsus, the martyr, not because he attached any importance to this role but because the transformation of the world of souls and minds is not timed by the man who is instrumental in this transformation. The timing is not man's business, not to his last hour is he authorized to know whether he is called forth to be a failure or a success. For that reason, Faraday made his successes "his outsides, and wore them like a raiment, carelessly." It was not his business to be successful. It was time for his science to become successful through him.

The actions of any real person, instead of a type, cannot depend on environmental consequences. When this person is a scientist, a mind in action, this detachment towards his own activities is especially needed. This is the contribution that has to be made by the power of the man's heart. The mind in action fails when it is not guided by the passive capacity of the soul. This balance between the mind's sky-aspiring activity and the soul's patience with the conditions of earthly bodies is the rare quality which lifts man to cultural efficiency. The pseudo-types serve their own desires and their self-chosen ends. It is often overlooked that as many people may fall in love with their being victims of life as with success. By taking it for granted that success is the only desire of man, we deprive ourselves of the means to study the laws of the good life.

Mankind never bestows the titles of classic or founder on a typical character. Types such as conqueror-victim, martyr and made man, failure-success, are on a level of pure nature, and for that reason void of that quality by which persons are interesting. The titles in the realm of historical creativity only go to people who to the very last keep the balance which we call freedom, the middle voice between infinite effort and infinite patience. The respiratory process between the two allows man to remain "in becoming", i.e., free, instead of making

37. *Timon of Athens*, III, 5.

himself into the product of his own pre-conceived, and for that reason merely typical will.

The common denominator of founder and classic is important, because it betrays the secret of their fecundity. They are fecund, the types are merely productive. The type is in his books, his inventions, his acts. The living historical person creates an army of disciples and followers in his image. The fecundity of Faraday and Paracelsus is far more important than their "output". Now this fecundity is practically non-existent in the mere type. The students of a typical scientist—a man with a big mind and an anaemic heart—are nearly always of lower ranks than he himself. It is well known that it is rare to find a rationalist producing students better than himself. The real person is an inversion of the type. The type acts objectively against and within the world, playing the world's game; inwardly, he is full of subjective desires, which he carefully conceals from the eyes of the world. The courage of classic and founder runs in the opposite direction: They frankly avow their heart's desire and are subjective within the world and society. Inside themselves, they are objective, because here, they detach themselves from their volition and take this not more seriously than any other objective element.

The historical person is passionately subjective in his relation to the world, and patiently objective to his heart's desires and his brain's reasons. Both transcend the type because they make use of man's freedom to be two in one: creator and creature, active and patient, planning and planned. When they act highly subjectively according to their calling, they, at the same time, submit to being timed objectively by mankind's destiny. By overcoming the common fear of the merely typical member of society, they invert the roles of courage and fear. Most men are courageous inwardly and indulge in all kinds of aspirations and desires privately; outwardly, we conform to all the requirements of society. A living soul, courageous against the world, fearful in her inner dealings with herself, leaves an impression not only on libraries but on other living beings as well. And apparently, this is the origin of their fecundity. Self-willed men, by admitting into their being the other portion of "being willed", cease to be types and become original personalities, classics or founders. Their contributions in matters of fact is one half of the historical role only: They themselves are contributed creators of a new type of man of whom they are first-born.

Two

THE SCIENCE OF BODIES AND THE APPEAL TO SOMEBODY

From the Richard Cabot Lectures on *The Future of the University*, in Cambridge, Mass., Winter 1938/39[1]

PRIMARY AND SECONDARY LIFE

Physics is losing its centennial rank in the hierarchy of our scientific universe. Physics ceases to be the spearhead of the marching army of thought. It will, of course, continue to work and to function. But it will not revolutionize the man who is not a physicist. The last three centuries had one main "Leitmotiv", one melody: mechanics, and that meant physics. Now, we hear the swelling of a new melody. And physics, although still with us, is denied its primary place by the physicists themselves. They hasten to disclaim any special leadership among all of us. This is the achievement of Albert Einstein. He removes his own science to a secondary plane. He cuts the umbilical cord that kept the growing body of science connected with its mother: natural philosophy, science of nature. Einstein says that physics produces its own concept of nature; it no longer owes it to any universal science. From

---

1. *Ed.* Richard Clarke Cabot (1868–1939) was an idealistic, wide-ranging physician of great distinction, based in Cambridge, Massachusetts, who found common cause with Rosenstock-Huessy on a number of issues. He sponsored a series of lectures by Rosenstock-Huessy at his home, on "The Privilege and Future Opportunity of the University."

its place as the first born of a numerous family of natural sciences, Einstein removes physics. He says that it is one science, without any authority for its use of the term "nature" over any other science.

He says that the "true" meaning of the physicists' conventions differs from the meaning generally implied. Now, before we ask more definitely what The Luther of physics has done to our colleges, let us ask what his removal of physics to another place proves, in itself, about our mental life. This question is of concern to educators. Because it tells us something about the rules for mental development in a living society.

Two methods of thinking delineate themselves: primary inspiration articulates meaning never before articulated, knowing that it has to be articulated for the first time. Secondary inspiration means re-inspiration by giving up deserted shells and going back to the "true" spirit of a dying incarnation.

The distinction between primary and secondary inspiration is of importance because it applies to any diagnosis of living processes. In the science of life, the distinction of primary and secondary processes is coming to the foreground in our days. Some biologists begin to think that the embryo and the mature man are not related as a mechanistic preparation—embryo to an adult final form—but as the free, original erection of a form to its later permanent functioning. Rudolf Ehrenberg compares embryology to the understanding of the artistic process of creation. The embryo is what the artist is in the realm of civilization. The embryo sifts, in a really more vital process, an infinite number of potentialities. His risks, his exposure, his originality is greater than those of the grown up.[2]

THE LUTHER OF PHYSICS: ALBERT EINSTEIN.

This is of some interest when we apply this to physics. Physics is not much more than 400 years old. It has been in the making in its artistic and embryonic phase. It seems to become a secondary racial process today.

Einstein is the Luther of modern physics because like Luther he sticks to the Bible of physics, mathematical language. One may think,

2. *Ed.* Rudolph Ehrenberg (1884–1969), German biologist, cousin of Franz Rosenzweig, member of a circle of friends that included Rosenstock-Huessy.

as we shall see soon, of a science of physics which does not use mathematics. Faraday was not well trained in mathematics. With Einstein, however, we are in the great tradition of physics which was formulated beautifully by Leonardo da Vinci: "No human inquiry can be called science unless it pursues its path through mathematical exposition and demonstration."[3] Einstein still talks the language of the physicist's Canaan. In mathematical language, he tries to speak the Truth about the physical universe. Also, he keeps certain naive basic dogmas of the old faith: there is one universe. This universe is a unity. This unity is a unity of recurrent possibilities, usually called laws of nature. It follows the line of least resistance. And the simpler solution is the more probable. Finally, the closed system of nature follows one course, towards entropy. That is to say, free energy is at the beginning, tied up; fixed energy prevails at the end. Less free energy is available all the time. All this is the universe of physics of the last 400 years. Nature is one system. In order to achieve the oneness, it is put between zero and infinity so that any experienced part of the universe is neither zero nor infinity. It is a directed system, running down like a clock which cannot be wound up a second time. It obeys the laws of probabilities.

Into this system, Einstein introduces the observer in his human time. The observer ceases to be a subject, a mastermind outside the space observed by him. He is made a part of it. Time is labelled the fourth dimension of space. This, although it has interested us before as poor logic, and will have to be discussed in the second lecture again, is of less significance at this juncture than the way in which Einstein deals with the observer. The objective world of physics, as objective as the visible church of 1500, is put on the stage of the observing individual. This individual, however, is a very purified individual, baptised with the baptism of science as much as Luther's individual. For, all these scientifically baptised individuals are completely equal: the difference between all observers in time and space may be neglected just as the multitude of Christian souls for Luther could be treated like one single soul. Luther simply took for granted that the differences of countries and centuries did not need to be overcome by any organic unity. And similarly, the body of scientists that has educated their disciples from generation to generation, this whole transcendent idealism and faith in physics, in objective space and in objective nature, is

3. *Trattato della Pittura*, Parte Prima

turned, by Einstein, into a convention. This agreement is said to be at the bottom of the whole scientific building.

This one convention, however, is only one out of many presuppositions in the science of nature. It is a much more complex historical heritage to believe in nature than [it is] to believe in God or to speak to man. There is, for instance, the presupposition of nothing. "Nothing" is the only unproved contention which makes all our positive statements possible. This is a bold assumption. Perhaps the idea of "Nothing" is the boldest assumption man can make. "Nothing" is not given in experience. Zero is a pure abstraction without concrete substratum from which it is abstracted. A line, a point, a circle in geometry are abstractions the concrete stimulus of which can be remembered. But zero? Yet, higher mathematics and physics could not exist without it. And infinity is also an irrational and amazing abstraction.

Both are imported into natural science and mathematics from quite external fields of thought. Zero is ultimately derived from man's experience of death. For the first Greeks on whom this notion dawned, it still seemed as if it ought not to be. They did not wish to call it "nothing", but "what ought not to be". Like the English word "lest", it deprecates. We build on "nothing" because nothingness must not exist; it stimulates us to transcend itself, to move away from it, to fill the vacant apace. And Infinity also was a notion which the majority of the Greeks refused to accept. The Greeks did attribute Infinity to their gods. The Heimarmene, fate, hung over the Gods as over man. Infinity entered our thinking from theology. Theology learned nothingness as man's mortality, and infinity as God's immortality.

That it actually penetrated into mathematics from man and God, is useful to remember. This fact explains, why at the moment when man's faith in God vanishes, physics require a new basis. It has borrowed, from theology and Humanism, the two notions which distinguish the concept of nature during the last four hundred years. The concept of nature as used by physics is untenable today, because the loan is withdrawn. The bank of theology and Humanism is bankrupt, The centres which made the notions of infinity and zero look "natural", can no longer give credit to physics. And we suddenly hear of limited space, of a finite universe as the last word of physics. Zero, now, is a convention based on neglecting the velocity of light. Zero is no longer real.

And so, Einstein, the Luther of modern physics, retreats into a building in which physicists dwell alone. More classic than the classical founders of his science, he cuts the tribe of scientists off from the common-sense tribe of man, son of man, child of nature, and child of god, all in one. Einstein restores physics by separating the axioms of physics from the rest of man. His science is a convention between experts, so benevolent and condescending logicians, physicists, and mathematicians tell us. They assume an air of disgust when laymen get excited over this principle of relativity. R[ichard] von Mises, in reviewing Einstein and Ilfeld's, "The Evolution of Physics", bristles with understatement: "Science is common-sense, used for remoter and rarer experiences. Physics has meaning for those experiences outside our daily horizon, etc., etc."[4] My dear and over-modest friends, your utterances reveal a deplorable lack of dignity. Formerly, infinity was true, and finiteness was untrue. Mind was absolutely stable; matter absolutely unstable. Copernicus was right and Ptolemy was wrong. Why was this so? Because the basis of your physics was laid, outside your department, by a general science called philosophy, the science of nature in general. And the clergy of this philosophy intended to deal, not with one special field of appearances, but with appearance. They never thought of these conventions as being conventions but as *binding* conventions. They deemed these conventions necessary and totalitarian. And they struggled violently to put them in the centre of every man's consciousness, as the leading principle of consciousness, of reason. Only yesterday, a colleague of mine wrote, in a book on God, that since the physicists had proved entropy, God vanished also in death through cold. Without exception, every field was subject to your conventions, because they were binding for physicists *and* everybody else, even a man investigating God Almighty.

As soon as you are just one group conforming to a standard, like cooks, shepherds, politicians, your science ceases to be of primary importance. It may drop out of our consciously cultivated horizon of first principles which we keep in store for unprecedented thinking. With the Reformation, religion ceased to lend itself to unprecedented problems. New problems then were tackled with non-religious tools of thought. For instance, natural law, mathematical jurisprudence, ethics *more geometrico*, replaced canon law, Roman jurisprudence and

4. "*Mass und Wert*" 11, (Zurich, 1938) 274.

Christian ethics. The general public is excited now by the principle of relativity, not because it is understandable, but because it frees us from the "general store" of natural science; we can't buy there any longer when we wish to deal with unprecedented problems in the future.

Unprecedented problems must be tackled by tools of primary vitality. Only the life-giving general ideas of an era have that character. Like the embryo, these ideas live exposed to myriads of potentialities; they are undetermined. For the living substance of humanity, a first principle like the Church in 1100, like Nature in 1500, has the same value that the plastic character of cells and tissue has for the embryo. These formative ideas can still respond to any unprecedented situation. And we re-establish our unity only when we are plunged into an unprecedented situation. It is then that we reclaim one tongue.

As a matter of course, such plasticity gets used up and lost. The infinity of potential responses is replaced by a circular response to those stimuli which have actually left their track on the plastic body during its growth. The repetitive response to relatively identical stimuli, we may call "functioning". In this sense, then, the concept of nature, in physics, begins to "function" after Einstein. My friend, the professor with his finite God, is obsolete after Einstein. He no longer has to take orders from physicists any more than from cooks. In our fairy tales, we hear of a time when cooking was so all important that the whole nation used the principles of cooking for every unprecedented event. Perhaps this is the reason why people began to cook their prisoners of war, too. A mathematical jurisprudence, or an ethics *more geometrico* strikes me as quite as absurd as cooking prisoners. Spinoza, to me, is a superstitious primitive, carrying over a first principle into an unprecedented problem, and worshipped for that reason, by all his contemporaries. And to borrow the motive of progress from nature and to speak of God's progress, compares to Spinoza's *more geometrico*.

Now, that all this should have happened, first the primary and universal significance of Nature in the Center of human conscience, and then its relegation, as a functioning partial thing, into the background, is inevitable. Like any living substance, a body of science uses up its potentialities; and that is its glory.

A science is a body of men sustaining the constant burden of doubt, halfway between ignorance and knowledge. A science is not the state of knowing. It is a perpetual restitution of an equilibrium between ignorance and knowledge. This is the reason why infinite progress in science is possible. A science must keep open toward ignorance and toward knowledge. It is an organized doubt; and the restitution of this doubt can go on as long as neither the unknown or the known part of the world is exactly the same in any given phase of the science.

PROGRESS OR VICIOUS CIRCLE?

If the research workers in any science should ever ask exactly the same question, and reject the same answer as they did once before, the progress of science would be imperiled. Since science aims not at isolated fact or data but at an attitude of people living between ignorance and knowledge, that attitude must be always new, otherwise life would go out of that science. We shall see that Einstein saved physics from this danger; and that other sciences are in the same danger now, only without a Luther to save them.

Before doing this, let us stop for a minute and weigh the physicists' assertions that nothing has changed, against our assertion that everything has changed. They can prove, by their publications, that they always have said that two and two make four, and that they never allowed witches or ghosts to take part in their procedures. And yet, the physicists' good conscience [today] has completely different content from their good conscience in 1500. Then, their good conscience consisted in having one word in common with all mankind. That word was nature. It is gone. The life-stream of humanity is feverishly searching for a new bed and groove in which to start for a new plastic and embryonic evolution of primary life and unprecedented experiences. And it is thrusting its consciousness forward in this new direction. And physicists now have a good conscience because they no longer have this common denominator with the primary intuitions of humanity. The sciences share the destiny of all organic life. You all know, from your personal experiences, of this transition from a formative stage to routine. We call routine what no longer occupies our imagination; it is so completely incorporated in us that our imagination is left free. We might describe this inner experience, with Rudolf Ehrenberg, as a retreat into a more remote interior of our own being. The part which was all and everything a year ago, and filled us completely, now dwells on the outskirts of our existence, while our heart and mind move else-where. Einstein restates physics, rejecting that universal philosophy of "a nature outside the observer" which had called physics, among other sciences, into being. I think of Einstein, as much as Luther, as a reactionary, or a last classic. Luther has no drop of secular thought in him although he prepared the world for it. Einstein has not one con-cept of a non-mathematical or non-physical character. This seems to be different in the case of other modern physicists. Planck testifies to a definite intrusion of thoughts which could not be thought under the government of that idea, so intimately connected with "Nature" with a capital N, the idea of a continuum. The adage *"Natura non facit saltus"* is well known. Planck abandons it. Similarly, we do find new ideas when physicists begin to transfer certain notions of living matter to dead matter. Certain scientists begin to talk of crystals, of electrons, as though they were organic substances. In other words: though Einstein still maintains the rigid notion of nature (bodies as mere matter and forces), the influence of biology begins to make itself felt in physics. This, to be sure, is only a dim foreshadowing of what will happen. The process will be reversed. More and more notions applying to living matter will be thrown into the gap opened by the fact that physics can no longer live on its analogy to God's infinity and man's mortality. Physics, in due time, will come under the protectorate of sociology, just as theology today is the handmaid of philosophy and science. That will take centuries, of course. At this point, I might interpret the place of mathematics as a social phenomenon. I might suggest that mathematics deals with those truths in which the time-difference be-tween the teacher's and the pupil's existence may be neglected safely. That explains why scientific education is relatively simple, the main crux, Time, making no trouble, here.

We may now understand better the history of philosophy dur-ing the last four hundred years. Physics and mathematics were at the bottom of the unrest and movement from Leonardo to Descartes—to Leibnitz and Spinoza—to Hume and Kant—to Bertrand Russell and Whitehead.

The truth of any living body of science is kept alive by struggle. The struggle by which physics came into being, was carried on in philosophy. Struggle in the schools of philosophy begot and fostered physics. In this sense, it may be said that physics has only just come

of age, in that it is fully emancipated from its parents, the two decisive schools of philosophy. Any surprise caused by this claim of philosophy to have begotten physics, will subside when we remember
that only fifty years ago, in every American college, the apparatus
of physics—as well as the globe used in geography, the ruler used in
mathematics, and the microscope—were labelled the apparatus of
philosophy. Philosophy, during the last 400 years, meant to think in
the light of nature. Philosophy was the wisdom of this world, with
the word "This" as much capitalized as "the Other World" had been
capitalized in theology. The two main forces in this science of "this"
world, then, were the empiricists and the system-builders. The empiricists, largely British, stressed the details to be discovered within
the new frame which held up, before the detached eye of reason, the
material world of space. The system-builders constantly repaired this
frame; they reworded again and again the implications of Nature with
a capital N. This school was mainly, but not altogether, represented by
continental philosophers.

We have here a significant division of labour within a living body.
It is not produced by "convention" as long as it is vital; it produces
itself, by moving thinkers to this front or to that with unconscious
passion. The word, division of labour, implies rational organization;
somebody divides the labour. In the life of philosophy, although labour was divided, nobody divided it among the mortal philosophers.
They found themselves challenged, every one of them, to take sides.
The risks, the exposure, the unprotectedness of the whole movement
seems to have invited champions, as knights in the Middle Ages took
up the cause of the unprotected orphan or bride. Only when we compare the process of philosophy to such immediately vital responses, is
it possible to understand the duel between the two European schools
of thought. You will remember our definition of a university as the
co-existence of different schools of thought in the same place at the
same time and in dealing with the same problem. Remember Paris,
Bologna, Salerno in their dualistic composition. We find here, in the
production of modern natural science, the same principle at work.
Instead of one city, all Europe is the scene of this struggle and dialogue. Europe is one city, so to speak, in which two schools of thought,
system-builders and empiricists, correspond by letters and academic
proceedings, and in corresponding among themselves, they really

respond to and represent the process of taking possession of *this world* for the humanity in which they live and think and write.

A third tradition, Jesuits and Lutherans, challenged the two schools as to their indebtedness to theology. It tried to admonish both Descartes and Bacon that their notion of Nature with a capital N was a historical creation, arrived at by an abstraction from man's state of nature as fallen, as complete nothingness. This third school was on the defensive. All other tones, like history and art, are merely overtones on this basic foundation of physics and mechanics. Today, Alfred Whitehead [1861–1947] tries to persuade his fellow scientists that their concept of nature is so void of reality that the old Greek cosmos, with its gods and men inside it, should replace it once more. Whitehead, as a restorer of a concept of nature (in the sense of cosmos) rightly comes at a moment when the delicate bonds between science and the "nature" of man and God in theology are finally used up and destroyed. His attempt to go back, is significant as a symptom. We are approaching the phase where science may lead to a circular movement.

For, if Whitehead could get us back to the monistic idea of a cosmos in which we should suddenly have to face not only physics but beauty, love, god, speech, all as elements of this world, then, indeed, the whole effort of the past 400 years would be partially cancelled out. Think, however, of our going back to the idea of a finite universe today: another danger of moving in a circle. About 1900, classical mechanics was in a dilemma which might have landed it in a blind alley or a circular movement: the fight between wave and atom, between matter and force, ceased to give results. As one physicist said: "Matter is victorious on Monday, Wednesday and Friday, and motion on Tuesday, Thursday, and Saturday." I think that, probably, Einstein has removed this danger of sterile repetition, by clamping the paradox of matter and motion in his "fourth" dimension of time.

THE CYCLE OF CLASSIC PHILOLOGY [OR] CIVILIZATION: WILAMOWITZ-MÖLLENDORF, 1450–1929

The fact that a science may derail is new to many. Let me show the danger of merely circular motion in another case. Physics has escaped the circular motion which would make progress impossible; philology as literary criticism of the classics finds itself in exactly the same danger at this moment. As you know, the Humanists, in strict parallelism to natural philosophy, discovered the natural world, which preceded Christianity. The "Nature" of science was paralleled by the "Nature" of antiquity. The natural world was infinite like God and the nature of the Greeks and Romans was perfect like Christianity. The yardstick for all human nature was recognized in "Classic Civilization" from 1450 to our days. Erasmus von Rotterdam exclaimed: *sancte Socrates*! And Socrates and Jesus were identified for the following centuries. Nietzsche embraced Socrates with his hatred because he hated Jesus. In murdering Socrates he killed the natural counterpart to Jesus. Modern college professors lecture on Socrates and Jesus in one breath. It is a mystery to me how they can do it. Plato takes the place of St. Paul in their scheme. As early as 1527, I find Erasmus saying that the fathers of the church were interesting only in so far as they repeated certain doctrines of Greek philosophy. In this way, the "Christian" texts were reduced to classical origins and sources. Physics traces everything to causes; it reduces. Literary criticism did exactly the same in the field of texts. From Erasmus, through Bentley and Wolff, to Gilbert Murray and Wilamowitz-Möllendorff and Werner Jaeger, philologists exercised the art of reducing texts to their origins. "Not Augustine first, but Plato already said; "—not Shakespeare, but Montaigne or Castiglione said", is the typical form of this science. The dissection of Homer is another famous case of reductionism. In vain that such a great mind as Ridgeway protested. When one reads Wilamowitz's last work on the Odyssey, with its violent destructionism, one rightly shudders at the tremendous powers of obsession. This great philologist had three chances to regain his freedom from circular psychosis during his youth. Three great men who withstood the temptation of mere reductionism, crossed his path. All three felt the European catastrophe of the World War nearing; and they knew that the whole game of Humanism which replaced Jesus by Socrates, and Paul by Plato, was up. The first was Nietzsche who resolutely turned to the Pre-Socratics and Dionysus, to the matrices and womb of Greek thought. Wilamowitz wrote a venomous pamphlet against him. The second was Erwin Rhode, the greatest philologist of his time, who probed into the religion of the Greeks (without the Erasmus-obsession of finding the purer and more natural Christianity among them). But Wilamowitz who (by the act of superposition), read into Plato the belief in God, Freedom, and Immortality, withstood Rhode's "Psyche" which investigated the lack of freedom, the ineluctable recurrence, the mythological bias of the Ancients. Finally, the great historian of antiquity, Jacob Burckhardt, tormented by the vision of the approaching downfall of the West, published his books on Constantine and on Greek civilization. Wilamowitz, this time, simply sneered. And after having denied the Lord three times, he went on for the rest of his life, as though driven by a demon, to reduce Homer. And with Gilbert Murray, he dominates the sunset of our era of a 'classical' civilization.

THE CYCLE OF BIBLICAL CRITICISM, 1770–1906: ALBERT SCHWEITZER

In secular philology he [Wilamowitz] did only what was done, with even greater zest, in the field of Biblical criticism. And here, the circular movement in the sense of a vicious circle, has been formulated by an insider thirty years ago. You all may have heard of Albert Schweitzer whose humanity led him to the Congo as a doctor who preached the Gospel to his patients on Sundays, but declined to be called a missionary. We, and the world, owe this new form of Albert Schweitzer to the crisis in Biblical criticism. Biblical criticism applied the methods used against the Fathers of the Church, to the New and Old Testament after 1770. It largely began with Reimarus.

In 1906, Albert Schweitzer wrote his "*Von Reimarus bis Wrede, Geschichte der Leben Jesu Forschung*". In this book, he showed that the circle was closed. Wrede, the last critic of the tradition on the life of Jesus, again asked the same questions of [as] Reimarus. Research had moved in a complete cycle. Every gospel, every letter of Paul, had come under scrutiny. A lost source, P., had replaced the authority of the gospels. The gospels had been moved into the second century of our era. The authors Luke and Matthew and Mark and, of course, poor John, had been stripped of their authorship. But one of these hypotheses contradicted the other. And in 1906 a great mind like Schweitzer could see that Christianity could not expect any light on the life of Christ from continuing this research. He studied Bach and medicine, and instead of studying the Life of Jesus, rediscovered the death of Christ, and went to the Congo. In him, you may assess the significance of the decision: progress or vicious circle. A human being that finds his mental activities enmeshed in a pagan rotation or the revolution of a cycle, will react by a violent jump. Our colleges cannot afford to let any science fall into the rut of circular movement, because that would destroy all loyalties in the students. Cynicism, violence, exodus, must be the soul's answer to the chances of such a silly game. Cycles are just beneath our humanity. They all belong to secondary and tertiary forms of life. Our mind was given us for keeping in touch with primary life, to reach out for the improvable (to use an important phrase of the biologist Rudolf Ehrenberg). All pre-scientific thought indeed, moves in cycles. Biblical criticism ceased to be a science when it went cyclical. I could show the same vicious cycle as the downfall of economics. I shall, however, stop here and not divulge the astrology of the business-cycle.

Let me make two points about this development because they will help you to see certain parallels in your own field. One is that Schweitzer's insight came thirty years before it was generally verified and incorporated. This lag between a person and a science seems to me important. In 1932, Chapman, the learned abbot of Downside, England, published a big volume which restored wholesale the original chronology of our gospels. The lost source P., this ghost of a century, disappeared again. Mark grew out of Matthew, and Luke grew out of both. At the same time, the Roman tradition that Peter founded the bishopric of Rome and was crucified there, was reaccepted as genuine by the scientific world. In scores of essays and dissertations, men did this inch by inch. When one of these men, again, had given in to one other point in our original tradition, I wrote him a letter, and asked at what speed he intended to continue this circular process. And why it was so important to give in by tidbits of one doctor's thesis after another, when the general principle and trend was so obvious.[5]

With Wilamowitz' death and with Nietzsche's devaluation of Socrates, the basis of our courses on "classical civilization" are gone. The idea of a purer "nature", of a humanity that is the true source and origin of Christianity, is gone forever today, when the noble savage attacks the very values which humanism as well as Christianity were

---

5. *Ed.* Rosenstock-Huessy speaks extensively about the gospels in *The Fruit of Our Lips*, ed. Raymond Huessy (Eugene, OR, 2021).

thought to embody. Nazism and Communism hurl their anathema against humanism and Christianity, and they quote the dark texts of Greece and Rome; they quote Frazer's *Golden Bough*, in their favour. The Humanists themselves cannot help falling in love with pre-socratic thought, pre-classic art like the Aeginetan reliefs, pre-platonic myth instead of Plato's ideas. The umbilical cord that connected classics and Christianity is cut. The very notion of the classic, then, is untenable as a general notion just as the notion of nature as a general hypothesis for our orientation is gone. The idea of classics and Nature gave our lives a clear place in the history of our race. They supplemented the existence of man in Church and State. To people who destroy Humanism and who don't even know of the Bible's existence, [who reflect] the alleged limbo of both, Plato, is uninteresting. And the same is true for Aristotle. In times of dogmatism and denominational precision, the father of definitions and of the syllogism was important. People today resent dogma and denominational precision. Why should they turn to their sponsor, Aristotle?

To sum up: Literature, literary criticism, linguistics, philology today lack their centennial fountainhead, shelter and roof. The world of classical nature in which the Renaissance believed as a kind of first edition of Christianity, collapses with "Nature".

The concomitants of a science of *nature*, in the sense of an uncorrupted lawful order—Greek and Latin and linguistic studies—must now look for a re-orientation. The study of Hebrew, Greek, and Latin will not keep their place unless they can find an absolutely new basis of existence. The classical world of an artistic "nature" borrowed from natural science its timeless existence in abstract space. And since science now knows that this abstraction from time is a mere abstraction for the study of extraterranean processes, the place of Greece and Romo in our college studies is unsettled. The philologists run around like mice seeking a loophole for protection and security in the new environment.

Mr. Einstein need not know what he has achieved. For, it was not he, indeed, who did it. He came when the times were fulfilled. However, the displacement of physics from its place as the first-born and very root of all the sciences cannot fail to involve all the departments which have lived on the assumption that Nature was a generality that reached from atom to Plato, from wave to music. That great Nature is gone which embraced everything except Revelation, and which came into being precisely with the purpose of rivaling Revelation.

Every normal American still holds this belief. And it is only among sober biologists that the downfall of the scientific hierarchy is seriously faced. I once more point to *Bios* I. (1934), by Adolf Meyer. As to the general lag and superstition of psychologists, historians, etc., I acutely remember James Breasted's last address, before the American Historical Association, on Social Idealism in Egypt and under F. D. Roosevelt. Finally, he said, the four thousand years of Revelation could be crossed out, and before and after that we might move in the refreshing air of purely natural idealism. This kindhearted anthropologist invited us to cancel out four thousand years of Jewish and Christian humbug. What can you expect of less kindhearted people? Breasted dogmatically believed that Nature was "better" than Revelation, and so he dismissed four thousand years of evolution, in search of his dogma.[6]

The physicists themselves suddenly disclaim the idea that their concept of nature has a meaning for everybody. Their's is a nature for physicists only. And that means that it no longer includes the nature of man, or even of other living beings or of literature (as "classical nature" did), or of language as the natural counterpart to revelation. *Life is unnatural, language is unnatural, literature is unnatural, man is unnatural.*

Our future line of demarcation will cut in between dead and living matter. And this is the decision, the cut which we have to make or lose our mental life. It is a matter of life and death for any teaching and instructing and investigating mind, to know the new boundaries or to add to the powers of darkness and death in his own activities.

Man, having put his head and mind once out of this man-made prison "nature", may go further; he may pull his whole being, soul and body, out of it, too. The "denaturalization of the mind" ([re Alexander A.] Jascalevich [*Three Conceptions of Mind: Their Bearing on the Denaturalization of the Mind in History, 1926*]) must be followed up by the denaturalization of creation. We shall have to denaturalize man more completely, to save him from decay, to save his life, his society, his humanity. For, modern savagery comes directly in the wake of the

---

6. *Ed.* Primarily a reference to James Breasted (1865–1935), *Dawn of Civilization* (1933)..

domination of Nature with a capital N. It is a scientific attempt to plunge man head over heels into that heartless, lifeless nature of the last 400 years. It is the final victory of the witches. Burned 400 and 300 years ago, these witches are unbridled victors today, with their black and white magic of the education racket, sterilization, drugs, surgical operations, Fascist-youth, guinea pigs, etc. etc. I am not speaking of the central scientific movement, but of the orgies performed in its suburbs, like psychology, or modern fiction, or Bolshevism or Naziism.

This conquest of man by his own idol, "nature", is not a return to nature so much as an advance *towards* nature as an Englishman has termed it wittily. The great God Nature has grown to higher and higher statures. Now Nature has become so big that man humbly offers himself as a bloody sacrifice to this idol of his own making. It is the indescribable attraction of mere grandeur which probably produced the mass slaughter of human victims in honour of Quetzalcohuatl by so kind a nation as the old Mexicans. Nature, in the form of race and proletariat, is getting human sacrifices again. The nickname, advance towards Nature, may convey to you this irresistible attraction emanating from the man-made idol "Nature" towards the modern masses. Our own makeshift, Nature with the capital N, is going to devour us, by denying us freedom, life, unity, creativity, peace.

By jumping onto the lap of his Buddha[,] Nature, man is spellbound by the big drum outside of him, and cuts his own throat. This drum of nationalism tells him that man has many natures, that you must eat others or be eaten by them. He is told that his heartbeat, his personal desire, his individual judgment, are nothing but blunders when compared to the nature of which he is a part. He is an artificially . produced African. And he is all this as a direct result of the final triumph of natural science over its rival, theology. And the physicists who now are afraid of this end of an era, and discount their own responsibility, are in the minority among their own clan. The scientific Nature-clan is still numerous among the scientists themselves. Lawrence J. Henderson [1878–1942], because his mind belongs to 1700, is driven step by step to intrude on man's nature in every department of Harvard. He sponsored [Vilfredo] Pareto [1848–1923]; he tried to have foreign Policy treated as the application of thermodynamic laws, he inspired an "anatomy" of revolutions.[7] Please look around you, and you will see your world filled with pre-Einstein naturalists.

Man has fabricated the notion "nature," himself. Man always transcends man-made notions. Man cannot belong to nature since there is no nature except by man's command. *To subjugate the maker of a notion to this notion always means to unmake him.* Either we unmake man, or he has to be believed and accepted as extranatural and unnatural. The denaturalization of life is the great historical achievement of the last 2000 years.

Language, logic, literature are expressions of this lack of entropism and naturalism. On the other hand, let us continue, by all means, to speak of the nature of things. And this reminds me that it should be possible to comprehend this whole diagnosis of the critical stage of many of our sciences in one person's grandiose attitude.

LEONARDO DA VINCI, THE FIRST INDEPENDENT LANDSCAPE, AUGUST, 2, 1473.

The nature of things was perhaps never presented to us better than by Leonardo da Vinci. He exclaimed, in the face of Nature: "By your law, you compel all effects to proceed along the shortest path from their causes." Leonardo, in fact, is the best sponsor of this notion: the nature of things. We have already quoted his paean on mathematical science. With an exclusiveness and purity which even today takes our breath, he emerged from amalgamate false natures into the artist, technician, scientist, mathematician of modern times. Not swerving to the left or to the right, not arguing with priests or lovers—unmarried, untonsured, unbound by anything else except his religious awe in the face of things—Leonardo, not Descartes, not Galileo, not Newton, and of course not that unspeakable featherweight Bacon, is himself the best man of the whole era. Truly, he is a child of nature. When he died, his pupil wrote: "*Tal uomo non e piu in podesta della natura.*" It is not in the power of nature to produce such a man a second time.

The pen of his disciple cannot help to form the word nature in this dirge. But what a strange phrase: "It is not in the power of nature. ... " In a way, we all know that this simply is true. As little as America

---

7. *Ed.* A reference to Crane Brinton, *The Anatomy of Revolution* (1938), a work by a Harvard historian with whom Rosenstock-Huessy had profound disagreements.

can be discovered by a second Columbus, so little is it in the power of nature to produce another Leonardo. In a way, however, we know that it is in the power of Nature to mix the elements so that she might stand up and say to all the world: this was a man, again and again. If we can be made to understand the twofold truth that Nature has unlimited possibilities, and that it is not in the power of nature to produce a second Leonardo, we may have understood the place of nature. And I think, Leonardo himself may help us.

In the year of our Lord 1473, on August 2, the first landscape was drawn by a human being, which was nothing but a landscape.[8] This landscape was drawn by Leonardo da Vinci at the age of twenty. It was his program, quite unknowingly. Before that, pictures used to go with poems (as in the East today), with legends and with narratives as symbols; and they were painted for their relation to man and God, to meaning and creed. This picture shows only valleys and hills, light and air, as a spacious sight. Nature is here, without supporting or decorating anything else. The background seems to exist for its own sake. These were the words that came to the lips of his last biographer, Antonina Vallentin: "*The background seems to exist for its own sake.*" I do not know of any more precise definition of natural science.

BACKGROUND "NATURE" VERSUS FOREGROUND "SOMEBODY"

Nature is the eternal background. The background contains all the possibilities that may come out of it at any moment: Leonardos, Napoleons, chameleons. In this sense, the background of nature will always be able to produce potential Leonardos. Now, the science of Nature is that bold enterprise of men during the last centuries to entertain the vision of *this background for its own sake.* That this is true, you may prove to yourself when we speak of the nature of a person, of a civilization, or of a group of people towards their betters. Whenever we relapse into the background, when a person dies, when a civilization collapses, when a science begins to move in a vicious circle, they all return into this background. Then, and only then, do we speak of the

8. *Ed.* The reference is to the earliest known drawing by Leonardo, signed and dated August 5, 1473, "Landscape Drawing for Santa Maria Della Neve," housed at the Uffizi Gallery in Florence. Rosenstock-Huessy misread the date on the drawing.

nature of a civilization or of a person. So often, in life, the only person who does not know the truth about his nature, is the man himself. Everybody else talks to everybody else about his nature; he never is told, from piety and respect; and so he dies of his own nature. Enemies render man the great service of telling him; and so he can let his nature die and rise again. The people who say behind our backs: "Yes, he is funny, you can't change him," simply condemn us to die. They treat us as nature; they push us into the background. The foreground is filled only with the impossible, the surprises, the improbable. The background contains the probable, the possible, the predictable. Scientists strive to sustain this background. They succumb whenever they leave the corresponding foreground, progressing, surprising, and achieving the utterly improbable. When scientists themselves try to become nature, background, their science collapses in a vicious circle. In the form of background, science falls back into the pre-scientific state.

Today, most sciences begin to move in a vicious circle; colleges begin to move in a vicious circle, on account of huge investment in buildings and machinery. The new million dollar machines in physics, easily may sound the death-knell of progress in physics. The background-science may exist for its own sake as long as something goes on in the foreground which is not for its own sake. As a product and child of nature, Leonardo is possible always. As a background, nature is inexhaustible. As the first painter of pure landscape for its own sake, Leonardo is the first man in history. Leonardo is the first citizen of the era of Nature. And this cannot be repeated. It is not in the power of Nature to send one of her children into our history at the same hour once more. We do not move in a circle. Life is open still. It has direction. It may push certain processes of secondary importance into circular motion, to get them out of our consciousness. But our consciousness must be filled with first-rate ideas, with life-saving ideas which are still unexploited and unrefuted. Here, life must go on as in the embryo, risky, plastic, unpredicted. Science must be forced out of its ruts in every decade. Man must survive his routine daily. A civilization must survive its habits in every generation. Things have to be done here once and forever.

Foreground exists for the sake of the whole background. All routine, all secondary forms of life, all the organs of our body, decay when they do not serve and are not keyed up again by the growth of new leaf, the bursting of one new blossom, by the one step into the unknown and into the improbable which we experience when we ask ourselves where our heart really is.

Einstein deprives the physicists of their privilege to move in the foreground of us all. The foreground, however, at this moment, is filled with an exanimate humanity which has been told to stare into the background only. This cult of the background of 400 years now asks its toll. The pedigree, the race, the environment, the laws of nature, the cycle, the curve, the background in education, the anamnesia, analysis, psychoanalysis, sources, origins, causes, reduction, is the dictionary of the modern person. And so we see him relapse into the limbo of the background.

Then, of course, when two people of different race marry, it is not called the founding of a new nation—which it is—but bastardizing. The revolt of the free is called maladjustment to the environment. The creation of a poem is just a contamination of sources. In looking into the background, we all become Orestes and Oedipus and Electra. In the background, causation is almighty. We call nature just one attitude of ours in which we forbid ourselves and the things of the background to have intercourse with each other. The background is the realm of objects. Anything put in the background ceases to have the right to talk to us, or to be talked to. It ceases to be a partner in our conversation. Objects are not conversant with subjects. Subjects converse with subjects. The whole attitude of natural science excludes the one commandment by which a foreground is created: that man must create subjects conversant with him. To say "nature" means to unmake subjects. To use the word "nature" is not the statement of a fact but the execution of a death warrant. In the world that is flooded with natural science, we ourselves are left exanimate on the battlefield.

Whom do you find in the foreground today? Children, maniacs, idiots, criminals. Only these seem to have the guts and the gusto to act in the limelight of a foreground. Decent people feel as if the background were the only decent place. In Chinese literature, the people vanish noiselessly through the back wall. In front, we listen either to the dummy Charlie McCarthy, or to Mr. Hitler. One is not alive, and the other does not speak, he shouts.

This has to be stopped. Fiat Lux! Let the curtain rise. Let us go out in search of the actor who is alive and who does not shout. One thing is certain: The background has a foreground whenever an actor has the courage to come out of the wings, to overcome his stage-fright, and to call another man's name. For recreating a foreground, a man articulates somebody's name. He does the only thing that nature does not do. He calls somebody into life.

Three

MAN MUST TEACH

Ca. 1940. Published in German in *Der Atem des Geistes* (1951), 95–166, and in *Die Sprache des Menschengeschlechts*, II (1964), 368–427.

SOCIOLOGY OF TEACHING AND AUGUSTINE

Aurelius Augustine was the last Latin Father who fought Greek and Roman Paganism. When he died, the Vandals were in Africa, and swiftly, the Roman Christians joined the battle of ancient civilization against the pre-city tribes. The new battlefront produced a union between Christian, Roman, and Greek elements. Soon, the monasteries became the archives of the whole ancient world.

When we read Augustine, we see for the last time the Church sharply separated from the ancient "World." Augustine had been a fine specimen of classic antiquity, and later he was bishop of Hippo for more than thirty years. In his "philosophical" student days, he had begotten a son—he had been seventeen years then—and now this son, Adeodatus was nearing the same age. Father and son were baptized on the same day. Legend has it that Ambrosius and Augustinus alternatingly intoned the

> Te Deum Laudamus
> Te Dominum confitemur
> Te aeternum Patrem omnis terra veneratur. . . .
> Sanctus, Sanctus, Sanctus Dominus Sabaoth. . . .

Augustine, the unlawful father, invoking the Holy Father of creation! The legend is magnificent. Unfortunately, it is silent about the son. On the other hand, Adeodatus did not live to see his father become a bishop.

Between Augustine's conversion and the son's death, these two people found themselves in a social situation for which neither Greek philosophy nor Christian doctrine had to offer much. For, here was physical relationship of a father to a son, born out of wedlock. By the act of conversion, this relationship was admitted to be based on sin. Here was religious comradeship, by simultaneous baptism of a thirty-three-year old father and his adolescent son. And there was the intellectual giant and roaring lion Augustine, and a young, inarticulate boy. Obviously, this situation was not harmonious. And no logic could harmonize it.

The father, however, seized on this bizarre situation. Adeodatus, at that moment, seems to have appeared to him as the new plantation the Lord had entrusted now to his passionate soul. And he decided to write a library, a collection of books or pamphlets for the benefit of Adeodatus.

This was in contrast to Christian usage. Flesh and blood shall not inherit the spirit, was a fundamental axiom of the Church; and the dogma of the Virgin Birth, the calling of Paul who had never met Jesus in the flesh, the institution of godfather and godmother, were only a few of the symptoms of this foundation. Sonship and discipleship, marriage and priesthood, were as strictly severed in the new zion as they had been identified in the old Israel. Augustine, then, was faced with the dilemma of becoming the Christian teacher of his carnal son.

Augustine saw the paradox of his task. He plunged right into the center of it. The De Magistro was the preamble of faith by which he tried to prove to himself that it could be done. The library never was written. But the De Magistro allows us to relive this peculiar station on his way through life on which the separation of flesh and spirit for which the Church stands, was to be reconciled. This is very modern. We are faced with exactly this issue. Can parents teach their children? We have broken up families on the one hand, and hear of Oedipus complexes on the other. Pressure from power seeking mothers, helplessness of wavering fathers, inarticulateness of all the members of the family on questions of faith, are mentioned to us daily.

The preamble of faith for any parent today must make answer to this: By what authority do I teach the children whom I have begotten physically and who are called my children legally? For, neither the physiological bond nor the legal relation explains the scope and limitations of a father's intellectual authority towards his son. The title De Magistro raises exactly this doubt. "Who is your teacher when I, your father, seem to teach you?" would be the full title.

In other words, the booklet tries to arbitrate between the roles of father, companion, hero, teacher, sinner, [of] which all five were united in Augustine and might well confuse the son. We do not know if this son was as hot-tempered as his father; if so, an early death might have saved him from an intolerable quandary. In his Confessions, the father perorates about the sin of begetting this son, and the innocence of this fruit of sin, a rather unsavoury declamation one might feel for a son to hear or even to sense. What a weight was laid on this son: the illegitimate child, the co-convert, the pupil, the follower, of a truly lion-like man.

If the waters of truth could pass through such a strange channel and yet be pure truth, this certainly deserved some clarification. The dissertation before us, then, is not an academic investigation on the merits of teaching in general, but a searching of hearts on the merits of this father's right to guide his son, in particular. Behind the dialogue, I cannot help feeling, looms the great question: Did Augustine have the right to have this son baptized with him? That had been done. The same step which in Augustine was the climax of a passionate life, had been taken by Adeodatus because he was this man's son.

This, then, is the significance of this booklet. It originated in a unique situation when Augustine paused between "world" and ecclesiastical hierarchy and came nearest to our own uninstitutional life-situation. Outwardly, the dialogue has been adjudicated to philosophy or to theology. But it belongs to a third type of literature. Of this third type, we usually only recognize biographical writings, letters or autobiographies. The De Magistro may draw our attention to the fact that these writings which are written to solve my own most personal problem, cover a wider field than merely autobiographical papers. Genuine sociological sources belong here, too. After all, a letter is part of a correspondence. And an important correspondence constitutes a fight, a wrestling between two souls. The correspondence between

Abelard and Heloise is not a philosophical or theological treatise, neither is it autobiographical. It is[,] because it is a correspondence, a sociological phenomenon. Strangely enough, sociology has shyed away from this phenomenon. A pair of lovers seemed perhaps too close to each other to be considered specimens of the social and group process. But the dialogue between Father and Son which is under our consideration now, cannot be classified correctly as long as we do not widen our categories. In this dialogue, father and son fight out the battle of fatherhood and sonhood. Now, what kind of literature is this? To call it personal, is quite as unsatisfactory as to call it biographical. For the two do not wish a personal solution: they are looking for a definite, for a true, and even for the scientific solution.

The exciting thing about the De Magistro is that it challenges our idea as though we could have a science of social affairs without this personal, biographical basis, at their root, or that we could enjoy letters and diaries, without the social truth and universal solution as their crown. We think for our personal salvation. And all social forms result from this fight for the salvation of persons. Of this, the De Magistro, is a telling example.

And this brings the book into sharp contrast to the usual literature on education. If it is true that it is written not by the famous professor of rhetorics Augustine nor by the bishop of Hippo, but by a father who felt uneasy about his prerogatives as a teacher, father, Christian, with relation to his son, student, fellow Christian, if it is true that he tried to find the truth and nothing but the truth not because he was in a scientific and detached mood, but because he was violently attached to his role in society—if, in other words, Augustine wrote this because he wanted to remain rooted and integrated, then it is possible that social science springs from personal bias and passion and belonging. Then, it is true that we do not teach others to do good[,] but because we, like Augustine, are compelled to teach by our own life's forces, even with the odds as in his case, against our qualification to act the teacher.

An objective adviser might have counselled Augustine to send his son to a public school or to an Episcopal school, and thereby to ease the strain put upon the younger man. Not so Augustine. Even he, who had sinned when begetting Adeodatus, wished and desired to teach this same son. Handicapped he well might consider himself. But teach he must. Teaching as an integral requirement of the right way of life, as a necessity even when the teaching is bad—that certainly strikes a new note in our discussions on education: *Man must teach*.

When we compare John Dewey's writings on education, and they are numerous and influential, we see the contrast. Never once does Dewey tell us why he must write his books or go on teaching. In discussing the foundations and underlying principles of education, the only regard he shows is for the little victims of our educational activities. The teacher is simply taken for granted. That he might be just as vitally affected as the student, injured, harmed, shellshocked, perfected, is no concern of most educational discussions.

Does this lack of reciprocity result from the idea that a teacher is a paid employee and that his salary is his reward? But if the pay is all he gets out of teaching, then teaching would be nothing in his life; and then, he can't be a good teacher. Nevertheless, educational theory modestly treats the sufferings of the teacher as pudenda not to be mentioned in good society. The parents, the pupils, the alumni, the public, are told why such and such a treatment will give the boy or girl the best possible education. A ware is sold. And this discredits our theories of education as advertising.

Any realistic approach would have to show how and why and that an adult can be induced to fool around with young people in this business of teaching and learning, some sporadically, some professionally, but all passionately.

The fact that John Rockefeller taught Sunday School all his life, that he did it, why he did it, how he did it, and if he should have done it, belongs as much in a scientific investigation as how and why and that John Doe should be taught the ABC. But the difference of these two questions is obvious. Question two can be debated in the absence of little John Doe who is too little to understand. The student's part in education lends itself to all kinds of abstractions, vague ideals, wonderful systems, statistics. But John Rockefeller or my first cousin, or an illegitimate father—their authority and qualification to teach piety and religion and history, must be debated in full view of their individual personalities and deficiencies and idiosyncrasies.

These people are real people, adult people, members not of the playgrounds of the schools, nay, taxpayers, adult social phenomena themselves. If the teacher's problem would form the basis of educational discussion, if we would ask: Can anybody teach? Must everybody teach? Should nobody teach? education suddenly would become politics and social science. But as it is, education is a humanistic and even humanitarian specialty since it is mere giving to somebody, with the teacher receiving a salary, in reward.

A sociological treatment of education must explain the lives of deans, scholars, assistants, janitors, alumni, college presidents just as much as of boys and girls.

Now, it would seem that Augustine was compelled to focus on the one point where all agencies involved in the educational process are fused. The overflow that is teaching and the influence that is learning, appeared to him as meaning one and the same energy. And man's relation to this energy stumped him.

Augustine is inexhaustible. He gave the Middle Ages and the Modern Times their clue. And now he seems to be able to fuse the two separated streams of our own consciousness, education and politics, into one new beginning. How might we call this third role of the man?

He gave the Middle Ages the basis of its axiom on faith and reason. Anselm took from Augustine his *Credo ut intelligam*, his metalogic [I believe in order to understand]. For a "metalogic" this truly may be called when I am informed for what to use my logic. Anselm used, and all the schoolmen followed him, the power of his logic to rethink all the experiences of man with his maker.

After this legacy of a "metalogic" had been squeezed dry, Augustine gave to the modern ages their metaphysics, through Luther and Descartes. The world of nature was dedemonized and as a created world lent itself to infinite rational inquiry. This complete severing of the ties between man and nature, mind and body, made possible the progress of science. In back of it is Augustine's metaphysics because Descartes could quote his doctrine that God was extramundane and man his rational agent with regard to the world if man purified his mind from all worldly attachment, if all scientists cooperated as one mind.

In both cases, of metalogic and metaphysics, Augustine placed the processes of logic and of physics, into a wider realm, into the life of the human soul. A certain soul, he taught, was capable of using its logic about God with impunity and usefully. A certain soul, he also said, was capable of using its physics about the world, without error and progressively. Under the condition that man loved his neighbor as himself, he could indeed know all these things without ending in witchcraft or gnostics. Hence all our science is universal and open as daylight since it is Augustinian.

Now, in his De Magistro, Augustine describes a third start. Here, he does not write the preamble to all reasoning about God by showing that he who makes any true statement, already must believe in the power by which we overcome our selfish interest and blind spots. He does not recommend detachment from the world before examining its facts. He writes the preamble for any member violently attached to his society, and trying to remain attached to it, despite the full use of his rational and critical faculties. In the search for a realistic sociology, we are beleaguered by abstract theories of education. Augustine says: that sociology must include the passions of the sociologist himself, his need for salvation. I the writer of this book, and you the reader, John Dewey despite his quest for impersonality, and all the students, both must receive functional satisfaction in a truthful order of education.

Augustine gives us the metaethics of utterance and communication. Before we can use our ethics of human relationships, we must be told *whose* life may use the functions and roles offered in these relations. Who is to become a father or a son, or a student? It is a certain being only which can escape unscathed from all these overwhelming formative influences and habits without being vitiated. He who enters into any correspondence, is to have certain qualities if his correspondence shall be worth anything.

Metalogic, Metaphysics, Metaethics—truly a giant the man from whom light may be derived three times, for three tasks, for theology in 1100, for philosophy in 1500, and now for sociology or social thought in general, in 1900.

THE DISTEMPORAN[E]ITY OF EDUCATION

It is not difficult to determine more closely the principle of Augustine's metaethics. And this will explain why he sponsors a science of society which puts education into the very center of all social processes and facts.

No thinker saw deeper into the riddle which "time" put before man, than Augustine. His remarks on "time" in the Confessions are rightly famous. But we will be able to quote many other usually neglected passages, on this subject.

Now a thinker who has something to say on the topic of "time" is ultramodern. The most energetic thinkers of our days, fret under this mystery of time. They are confounded by the fact that the mind may be thought of as observing the bodies in space, but that this same mind takes time to function at all. True enough that the mind observes the facts of the world of space. But we seem to be unable to observe time since our own thinking takes time. The subject of the thinker is subject to the time stream, is conditioned by time. But how can that which is conditioned by some force, ever be empowered to understand this same force? If we are the products of our time, we shall never know this same time as we may know a fact of outside nature.

Thinking takes time, education takes time. We send our children to school for a dozen of years. But modern scepticism has dissected time and found that it consists of disconnected atoms, seconds. The largest school of thought in this country teaches that time knows of past and future only, that the present is of a razor-blade short-livedness, and that when we speak of "the present period," we are handling a fiction. They call all usage of a present in this larger sense, a "specious present," a fictitious unit of time. An hour in the classroom, a war, a revolution in which we find ourselves, are all fictitious, according to these logicians. And logic seems to be on their side.

But if this is so, then farewell to education. If a class consists of disconnected split seconds, education is impossible. For, all education plans a curriculum of years as though time stood still, in a certain sense.

Augustine suffered from this contradiction. And he pointed out the direction in which the solution may be found. And the snobbery of the modern sceptic which declares the present as not existent and believes in past and future only, melts like a snowflake before his scrutiny.

The De Magistro would be too fragmentary if we would not read it within the framework of Augustine's philosophy of time. And vice versa, our reasoning about time receives a sound basis, if we fathom the depth of the fact that our own thinking about time takes time.

For nowhere is this more in evidence than in the classroom of educational institutions. Teaching is not peripherical for a science of time because it makes transparent the fact that thinking takes time. In any act of teaching, time is of the essence.

So much is this the case, that time appears there in at least three qualifications at once. First, there is the schedule of the whole curriculum, second, there are two kinds of people, one older, the other younger, both with a time of their own, and yet thrown together into this identical schedule.

It seems that we have here in a nutshell the time-compound of all social relations. The teacher and the student are not contemporaries; yet they are synchronized. Hence two "times," two lifetimes, seem to be able to join. Without this basic belief, teaching would be impossible. Whatever else teaching may be, if we restrict its aspect to the purely chronological skeleton in it, it always shows two people at least one of which is, with regard to the subject matter taught, ahead of the other. Now to be ahead is here simply an expression for the teacher's pre-acquaintance with the matter. Five minutes earlier than his student, he must have come to know it at least. Whereas in all other cases, the difference between old and young may be glossed over or forgotten, in teaching, this discrepancy is made the cornerstone of the whole process. Here, a difference in time is necessary to make the flow of experience possible.

Teaching is based on a succession in time, willy nilly. And the reason why the teacher should give his time to a young brat and why the young should place his faith in an old ass, remain to be explained.

Augustine does exactly this. He sees that a social itinerary must link together the young and the old, the primitive and the educated.

Indeed, in teaching, the social system reveals itself to be based on a harmony of innumerable times. People of different age are made to coexist. But different age also means different ideas, different interest[s], different outlook, different taste, different beliefs. And yet teaching? Yet a flow of light from the representative of one time to the representative of another? This is not an academic question. How many parents actually did say, during the last decades, that the times are so different that we can teach little to our children?

Yet, as long as anything is taught, the collision between various times and their different truths is considered to be superable. The relative character of all differences in time-truths is therefore the basis of all teaching. But this means that all teaching makes definite assumptions about our relation to time and submersion in it.

And this is indeed true.

The difference in age between co-workers may be accidental; the time difference between teacher and pupil exists by establishment. They are, therefore, distemporaries, not contemporaries. Two times exist of which one is embodied by the teacher, the other by the pupil. In learning, in teaching, in education, the miracle is achieved of bringing both together in a third time. This bridge is called the present.

Now, I cannot find that anyone except Augustine has pondered over this situation. I have looked up, for the purpose of verifying this proposition, a long list of books on ethics, medieval and modern. Nowhere did I find that they saw a problem of the first order in the time abyss between teacher and pupil. Here, the darkest division of man stares us in the face. And our handbooks on ethics deal with justice and property and crime and labour and government. Education comes as an appendix, with all the optimistic colours of the easiest part of the ethical system. And the teacher in us is mentioned nowhere, with his rights.

Augustine saw that all our troubles spring from the educational task. For, to him, we small men are expected to form together one great man through the ages. From Adam to the end of times, man is one. The ages die. The generations die; the individual passes through at least seven ages during his little life. And yet the spirit's bloodstream survives every one age. For this grandiose task the different times and ages of man must be made co-existent although every one of them only lasts a short time. Augustine says in *De genesi ad Manichaeos* I, 43, "The age of the mature man corresponds to the fifth day of creation when fishes and birds are created. Hence, this man must teach, pervading the air like a bird, with the winged words of celestial teaching. And he breaks through the waves of time, like a whale, with the power of contempt. His students, on the other hand, and their *aetas* [i.e., age], compare to the second and third day of creation. For, whereas as infants, they are bathing in the undivided light of the first day, the boy and girl begin to remember and to distinguish. And the very first distinction is between heaven and earth, high and low, carnal and spiritual. In this way, the ages may imitate eternity by their co-existence."[1]

It is, therefore, in line with St. Augustine to put the process of teaching in the centre of all sociology. This is the only important distinction between a Christian sociology that is based on the word, and a naturalistic sociology. Usually, people derive the authority of a teacher merely from his expert knowledge. When we do this—and St. Thomas does it—we fall into the abyss of departmentalisation. When people deduce the right to teach from the "State," they fall into the abyss of propaganda and lying. It is only when teaching is based on no other, external or logical, process outside itself, when education is recognized as an original and irreducible situation between two souls that we escape the hell of -isms, of inquisition and propaganda. We all need an answer to the simple question: How can people who are not contemporaries live together successfully? And Augustine's answer is: They succeed if they admit that they form a succession, if they affirm their quality of belonging to different times. If the time difference is admitted, they may build a bridge across the times, in corresponding acts. By these acts, that which is called "the present" is produced. The present is not a given data of nature but a fruit of social efforts.

The teacher's unrelated lifetime before he acts the teacher and the student's unrelated lifetime before he becomes this teacher's student know of no present except as the razor blade between past and present. When the two converse, the man A by acting the teacher, concedes that he represents the past, and the man B by acting the student, volunteers to represent the future, between them. And by taking upon themselves these two roles, a present emerges which stands above the past and the future as their common ground.

---

1. *Tempora fabricantur et ordinantur aeternitatem imitantia. Orbes temporum numerosa successione quasi carmini universitatis associant.* (The times are manufactured and ordained as to imitate eternity. The periods of the times by numerous succession organise themselves as parts of the song of the whole.) De Musica, Migne, *Patrologia Latina, Opera Augustini* I, 1178.

ANALYSIS OF THE TEXT

In two chapters, we have dealt with the situation of the dialogue between Augustine and Adeodatus, and with the problems of time and education which it raises and against which it should be pitted.

We now proceed to an analysis of the text.

The text consists of fourteen chapters. We shall sum them up, one after the other.

1. By speech, albeit prayer, song, or teaching proper, we cause the very things to come into the mind of which the words are signs.

2. In commenting on poetry, we are expounding words with words, signs well known by signs equally well known.

3. In as far as man asks questions by means of words, he usually must put up with words as his reply. He may, however, get his answer through other signs or gestures, or the act itself may be performed.

4. A sign may point to things or to other signs. The word "noun" or "conjunction" points to signs; horse and river point to realities.

5. Every sign is both: sign and meaningful. Words are signs with regard to the ear, and meaningful nouns with regards to the soul. Any word (for instance: "if," "because") can be used as the subject of a sentence, i.e. as a noun.

6. Some signs signify themselves like the word "word." Others are reciprocal like vocabula and nomina. Some signs are synonymous. Words from different languages differ acoustically only.

7. Adeodatus sums up:

    All speech is teaching.
    Words are signs.
    Signs need not be words.
    Acts may be shown without a sign.

8. Augustine himself sees these points clearer now[,] *quam cum ea inquirendo ac disserendo de nescio quibus latebris ambo erueremus.* (Tourscher: by questioning and arguing we both were

drawing them from some unknown obscurity; Leckie: we un-
earthed them from unknown hiding places.)[2]

The goal of this discussion is difficult to explain. Adeodatus
may either consider this to be a game or expect some small result
or he may become impatient because he is hoping for a big result.
Augustine *although playing* is not aiming at a toy thing: "On the
other hand, it may seem rather ridiculous when I pretend that
it is some blessed and eternal life to which I wish to be led with
you here under the guidance of God, and that is to say, of truth,
namely by some steps that will be appropriate to our poor gait.
For, I have entered upon this highroad not by studying the objects
that we signify, but their signs only. Yet, this *prelude* exercises the
very energies by which the warmth and light of the region of the
blessed life may be not just forborne but truly loved."

* * * *

The two syllables ho-mo may mean a real being, or these two phonetic
fragments. Generally, the presumption is in favour of the reality of
which the word is a sign. When we ask about the word as a word only,
we should qualify our question. It is legitimate to answer an unquali-
fied question as though the real thing was the object of the question.
Sophists are abusing this righteous attitude.

9. A sign may be equally or more valuable than the reality sig-
    nified. But our cognition of the sign is less precious than our
    cognition of the reality signified. Examples are "filth" and "vice."

10. The assumption in chapter III and VII that certain acts like
    walking are self-explaining, is refuted. Result: Nothing is taught
    without symbols. Adeodatus feels uneasy. Augustine, in fact,
    turns the tables now and shows that everything under the sun
    may teach us without the use of signs. We even understand new
    words only when we see the object which they signify.

11. "To give the maximum of credit to words, words challenge us
    to seek reality." We may and shall believe words. Understand-
    ing, however, should follow as frequently as possible. And

2. *Ed.* Francis E. Tourscher, *The Philosophy of Teaching. . . . A Translation of Saint
Augustine's De Magistro* (1924); George G. Leckie, translator of *De Magistro*, 1938.

understanding is not produced by words. It is not even achieved by the speaker although his words may challenge us. *Tantum cuique panditur quantum capere propter propriam sive malam sive bonam voluntatem potest.* (Leckie: There is revealed to each one as much as he can apprehend through his will according as it is more perfect or less perfect. Tourscher: It is opened out so far to each one as each one is capable to grasp by reason of a good or a bad habit of life.) See our criticism on page 17.[3]

12. Sensations and mental perceptions are the two classes of our perceptions. Sensations never are replaceable through words of others, except on faith. In a case of mere belief, nothing is learned. The same is true of mental processes. "The auditor whom I tell that I saw a flying man, will answer: 'I don't believe you.' In the same way, he will deny the spiritual truth which he is not fit to know." Any auditor will either accept on faith, or deny, or consent by his own spontaneous testimony. In no case, then, will he have learned, properly speaking.

13. The listener is the speaker's judge, or at least, he is judging his speech. The speaker may quote texts in an attempt to refute them, and the listener still may approve of this very quotation. Sometimes, it is true, we succeed in speaking our minds. However, we are talked to by as many lying people as by truthful men. Besides, by inattentive talking, slips of the tongue, etc., any number of quarrels and misunderstandings may be produced.

14. Nobody sends his children to school to let them think the teacher's ideas. They ought to get the objective knowledge. This they only learn by spontaneous consideration inside themselves. That we should call the man who speaks to us, "magister," springs from the fact that no time seems to intervene between the moment of his speaking and the moment of our cognition. Because this time element is overlooked, the students think that what they learn from the interior truth, has been learned from the external admonisher.

The general usefulness of words which, well considered, is not small, we shall investigate elsewhere. Here, however, I wish to restrict

---

3. *Ed.* See p. 121 in this edition below.

their importance. *I only have admonished you.* We should not only believe but also understand why it is written with divine authority that nobody is our master on earth since one master is in heaven. Matthew XXIII, 8: "but be not ye called Rabbi; for one is your master, even Christ; and all ye are brethren. 9. And call no man your father upon the earth; for one is your father which is in heaven. 10. Neither be ye called masters: for one is your master, even Christ."

With all my questions, with all your answers, you have not learned from me. Confirm me, Adeodatus. And Adeodatus affirms:

Words from outside are admonitions. He only teaches that dwells inside. And I have experienced this during your talk which I have enjoyed. All doubts were dissolved by the inner "oraculum."

Some discarded digressions in De Magistro:
Ch. I. What is the intention and value of music?
II. "Nothing" is a difficult problem.
III. The words of prayer are not the essence of prayer; still, they have their proper social function.
IV. It remains unsolved how a term like 'ex' should be defined.
IX. A thing that serves another object need not be inferior to that object, Adeodatus thinks. Augustinus holds the opposite view.
XIV. The positive usefulness of words is not to be discussed in this dialogue.

REPENTANCE FOR A SOCIAL SITUATION.

The dialogue deals first with the meaning of speech, and then with the origin of truth for the boy who is spoken to. The dialogue takes place between father and son after they have left Italy and wish to establish themselves as baptised Christians in Africa again. As a dialogue, it still preserves the technique of that academic life that Augustine and his friends, including the son, had led together in Italy. On the other hand, this is the only piece in which father and son are on their own resources, without anybody else. The instinctive loyalty to the form of production that the life in Italy had asked for, is obvious; on the other hand, the death of Adeodatus left this dialogue as a mere fragment.

Augustine's life in Africa soon followed a new pattern, of public and ecclesiastical character. Thus, the De Magistro is the obituary of a boy who must have been full of life and wit. And the boy no longer was a boy. He was seventeen; at that very age, Augustine himself had begotten Adeodatus! Adeodatus is on the verge of independence and maturity.

At the end of the last chapter, Augustine hints at the situation in which the dialogue is written. It is meant to be the forerunner of more to come. The intervening death of Adeodatus has kept from us the sequence of De Magistro. And what does Augustine plan as a sequence? This is very important to know when we wish to interpret that what we have and what is a fragment only of what we would have without the loss of the son. For, if Augustine announces what he is going to do later, we may be sure that he does not think to have given us this same thing in De Magistro. And this indeed is the case. Augustine promises to write on the usefulness of words "which when rightly considered is not small." The De Magistro *shows how the use of words should be "rightly" considered*, without being itself the positive treatment of this usage. The De Magistro is not concerned with the positive teaching of grammar, speech, etc. as the modern significationists would like to find. "Foundations" are laid. Today, the use of the word "foundations" is handled so loosely that the meaning of this word is forgotten. Mr. Leckie thinks that the first chapters of De Magistro contain Augustine's final ideas on the subject. The whole dialogue, however, moves away from these introductory chapters. And any "foundation" has to do so. Why is that so? Foundations wish to get away from a surface that is unable to carry a building. We go against the surface and away from the surface not by building a skyscraper, but by excavating the ground when we lay foundations.

In the Liberal Arts community, in the situation existing between Augustine and Adeodatus, between any teacher and any student, there is danger, there is abuse. The foundation must be laid anew for the rebirth of the school. Everything will sound in the reborn school differently from what it now seems to be in the unregenerated school. Hence, all the grammatical and rhetorical arguments in the first part of De Magistro only serve the purpose of describing the processes in the unregenerated environment without passing any judgment on their final value. The purely descriptive character of the first part of De

Magistro as a specimen of what people use to talk in schools removes our book from the Platonic pattern. It is not imitative of a Platonic dialogue. A social and scholastic situation is described and enacted so that it may do repentance and be lifted upon new foundations. The first half might be compared to Abraham's attempt of sacrificing Isaac. We are told this because at the end, Abraham instead sacrifices his own will. In the same way, the first half is narrated by Augustine so that it may be jettisoned in the second! The dialogue is a biographical event in the life of the two partners. Thought is political; this dialogue does not dwell in the realm of theory; it is an act within the practical life of Augustine and Adeodatus. [Jean] Guitton has some very beautiful remarks on this difference between Greek and Christian thought; he says, "The unsurmountable abyss between Greek and Christian thought is the Christian rehabilitation of the unique and temporal event. The moral order is general and abstract to every philosophical or Greek mind. In Christianity the time of every human existence receives a superior quality in its smallest fragments."[4] One of these smallest fragments is the hour between 8 and 9 in which I am writing this essay or the classroom lecture in which logic is taught. By the Greek mind, or as we call this mentality today not quite as sharply, by the academic mind, this fact is ignored, a lecture was thought to be a theoretical display of thought. Hence, it would seem that in the classroom, the events, the ideas, the people that the teacher mentions enter into a merely Platonic realm of ideas. In imitating the ancients, the classroom, the teacher and the students feign to have timeless minds. On these minds, the events, people, ideas mentioned in class leave an imprint, as a movie does on our imagination, with the movie moving and ourselves sitting unmoved. In the dialogue De Magistro, this academic atmosphere and disposition disappear. Here, we have no difference between theory and practice. Augustine and Adeodatus think out their salvation as chapter 8 clearly says. The dialogue is not academic but biographical for both. It is a social struggle.

The whole dialogue and especially the break in chapter 8 remain ununderstandable as long as we think in academic terms of a difference between theory and practice. However, a dualism is here too; the book is obviously made up out of two parts. Only, this is another dualism, the only dualism admitted by a Christian community. It

4. *Le Temps et L'Éternité chez Plotin et Saint Augustin*, Paris 1933, p. 359.

is the dualism between play and seriousness. This dualism is at the bottom of the dialogue, and Augustine says so himself. We never are "academic," but we alternate between play and struggle.

The dualism of one non-committal and one definite part divides the dialogue right in the middle into two septenaries of chapters. Out of fourteen the whole consists. In chapter seven, Adeodatus sums up the results of the first six chapters: "What do we do when we speak?"

Many sides of this question have been mentioned and left unsolved. They are listed at the end of the summary as unfinished digressions. The father has freely avowed his ignorance in some cases; and the son has been as often right against the father as the father has been against the son. They have cracked a number of jokes. For instance, in discussing the word "nothing," they discover that it is a wonderful sport for sophisms on "nothing" when this alternatively may mean the word "nothing," or the difficult concept "nothing." Augustine gives up after a while, jokingly: Come on lest "Nothing" us delay.

The whole first part is remarkable for its good humour and its poor results. And no wonder. *For, we learn in chapter 8 that this was a play, a prelude, and an exercise only.* And to prove that he means what he says, all the dearly bought results of part One are refuted or given up in part Two. At the end, we do not know what is true in this respect; and what is more, we do not care. What has happened? Augustine says explicitly that he wishes to lead both into a quest for the good and blessed life; however, he has taken an unusual start. Mostly, when a moral issue is involved, we plunge directly into the material problem involved. Instead, this time, the conversation begins with a reflection on the means of discussion, of speech and the signs used in speech. These very signs may be taken too seriously. And that is why Augustine wanted them to be shown up in their relative importance. The first half of the dialogue plays with the unimportant; the second is seriously concentrating on the essence.

Some of the modern Augustinians will dislike the idea of dismissing a part of the discussion as less important. To the logicians, a difference in importance is a foreign idea. They are serious all the time; and so they become ponderous. I suggest that just this has happened to Mr. Leckie. The conditions of play and work are nearly unknown today to the philosopher.[5] Yet, it is a fact that people who

5. See my *Soziologie* [. . .] 925 on these two points.

live together must play and work together, both. We play together in our state of innocence. We must work together for our sins. 75 years ago, Horace Bushnell wrote an essay on play and work in which he said that play was the normal thing, and work should be lifted up to the level of play.[6] And the church holds that the liturgy is a play of humanity in the face of their Father. In Heaven and so far as we are in heaven, we play; on earth, and in so far as we must work out our salvation, we struggle. The dualism that divides human activities, is the dualism between play and struggle. The difference between theory and practice is a fallacy. Thought is struggle as much as any other doing. Of course when we compare leisurely thought, irresponsible talk on one side, and responsible labour and toil on the other, the division between mere theory and realistic practice is very tempting. We are misled by the fact that in this case the act of thought is a play, the act of our hands is serious. Serious thought and wilful practice would be divided the other way round: the practice of the player is quite irresponsible, purely "theoretical," the thought of the doctor who tries a diagnosis, is strictly responsible, hence the most real practice. Let us replace the futile division: theory versus practice, by the realistic: play versus struggle.

In allowing Adeodatus first to play with him, Augustine prepares for the full warmth and light of that region where the blessed life is lived. Today when we work with one kind of people and play with another, our best thoughts remain our private property. Why has everybody today a private religion only? Because we cannot find the truth together when we do not play together. For that reason, we find little truth together; most truth that we find remains our private affair. The dialogue itself, in its method, is a specimen of how people may find the truth together.

By this method, Augustine is able to lift Adeodatus from one level of thought to another. This shift of level is the real goal of education. As long as people think of teaching merely as the instruction of facts, this shift in level is overlooked or even denied. Many teachers would say that we move on one and the same level during a lecture. That this is not true is proved by the simple fact that even they cannot help cracking a joke once in a while. If they would analyse the impact of this one little fact they would face the real educational mystery which

6. *Ed.* Horace Bushnell (1802–1876), American Congregationalist minister.

is that man meets his fellow man only when he meets him on different levels. This is not a logical proposition; and it is not a psychological proposition. It is a social and historical phenomenon. And this is Augustine's problem all through the De Magistro. *The student plays, the teacher struggles with the truth.*

The student is faced by a question in the classroom which to him has not yet become personal. For, we anticipate life's experiences by going to school and by learning from others. Hence, the things to be learned even by the best and most eager student, are faced from afar, and this gives the student an attitude towards these questions as though he might toy with them. He, every adolescent, plays with ideas. As a friend of mine said to me: "Never take a man up on what he has thought before he was thirty." Hence, the play situation is represented by the student. The teacher, it need hardly be said, is the more entitled to the function of teaching, the more he has wrestled with the question in dead earnest. He may not struggle any more; but at one time, he must have struggled with the truth he is going to propound.

The past though past is serious. The future though approaching can still be played with. And the student's playing with *ideas*, compared to the teacher's *convictions*, compare like regular current and power current. We need a transformer, to bring the truth from the form of conviction to the form of play. Otherwise, it will not be accessible to the student. Hence, Augustine did play with Adeodatus first, and was quite willing to jettison part of his truth as having not much weight.

But this is not the whole process of teaching. For, the student must be made aware that the teacher is in earnest and that he, too, one day, will have to be in earnest. The transformer must work in the other direction, too. The playboy-attitude must be stepped up to seriousness. In the same manner in which the teacher shifts from his own plane to the student's level, the student will have to move from his lukewarm and aloof attitude to eagerness and enthusiasm. He must be aroused to two acts. One, he must recognize and respect that the teacher is reporting a struggle, not a play with truth. The other, he must follow him into this struggle himself. Good teaching begins with a joke and ends with a challenge. They represent the two levels which wait to be equalized by the transformer called teaching or education.

To degrade teaching into puerility or to sublimate it into crusading, are the two dangers of teaching. Play and struggle, low voltage and high voltage, shall be equalized. When they are, all that which education can do, has been done. Both partners enter into this process as completely as electricities enter the transformer. Neither the teacher nor the student are master of this free process. It has an elementary character. It may succeed or it may miscarry. As long as we overlook this aspect of teaching, the relation of struggle and play, of conviction and idea, we may think of education as a safe trade in which some ware changes hands. And our recipes on "techniques" betray this evaluation of education as a thing which can be mastered by the teacher and of learning as a process to be mastered by the will of the student. And it is true, instruction can be drilled in by relatively safe methods. Knowledge and information can be imparted by sound techniques. However, nothing of importance about man himself can be transmitted without the full investment of two real lives in a situation which is and remains risky. The more important the topic of teaching, the more risk is involved. The struggle and seriousness may be misunderstood, the jokes of the teacher may be misinterpreted. And when Hegel said: "I had one student who understood me, and he misunderstood me," he had the courage to crack a joke which was much more than a joke. He who has never been misunderstood, may be sure that that which he had to say was not important.

The truth, conveyed by play and struggle both, must emerge beyond these two forms in which the student and the teacher conceive of it.

And this is the text of the second half of the dialogue. Since teaching miscarries so often, Augustine tries to eliminate some frequent causes of the miscarriage. The simple fact that we get involved into any kind of conversation and social intercourse, invariably exposes us to the danger of misunderstanding and of being misunderstood. The signs and words used in speech, learning, teaching, seem to be "owned" by the interlocutors. We credit them with their meaning. We view them as the masters of the situation. The term "original sin" is not used by Augustine with regard to this situation. And it is well known that he never was able to solve the mystery of this concept of original sin to his own satisfaction. But the situation in which we find ourselves by conversing, is not far distant from the dilemma which the church described by this term. We are near it when we see Augustine describe the indecisive and arbitrary plays of mere sagacity and dialectics and how they becloud the moral issue which props up in every conversation. As soon as we are unaware of the *risk* and consider the partners of a conversation as its overlords and not as elements in an unforeseeable risky event, of which they are mere subjects subjected to undergoing it, we attribute to them a power which they do not have. What happens when we have played together? What does it mean when we become serious? Does it mean that we become thinking machines? Understand the decisive turn by which the teacher's role is transposed from a rational, logical, pragmatic, scientific, and scholastic role into the realm where it really belongs and within which it becomes clear that every man must teach. And why human beings are by nature obliged and authorized to teach as much as we assume that every child should take the opportunity to learn.

The teacher is stripped of his logical togs. He may be a great scholar or an expert or a logician or a scientist. But *in the act of teaching, he does not function in this capacity of a "mind" or intellect.* In the process of teaching he gets involved because he has a soul. In Augustine's metaethics, it becomes obvious that the teacher must be satisfied with an ethical role. Any such role is enacted not by the mind in us or by the intellect but by the little something without which the modern mind would like to explain education and teaching. William James thought that our rational explanation of the universe did not stand in need of this little something. And modern psychology and John Dewey's philosophy dismiss it with a shrug of their shoulders. William James, at least, admitted that the little thing might have to be allowed in again if a champion could be found who could show some pragmatic significance for it. Now, Augustine is this champion of the exercised [excised?] term "soul" because teaching cannot be explained if the teacher has no soul.[7]

For, the teacher is torn between his duties to the truth and his love for the pupil. AND WE CALL "SOUL" THE POWER WHICH CAN TOWER OVER OUR TORNTOPIECESHOOD BETWEEN CONTRADICTORY TENDENCIES IN US. The soul is the power to forbear conflict. The conflict which the teacher takes upon himself lies

---

7. *Ed.* See Rosenstock-Huessy, "The Soul of William James," in *I Am an Impure Thinker* (1970, 2013).

between his thought in his own time and the survival of this thought beyond his own time.

What is the situation? The man of good will learns, Augustine says (chapter 11). The boy of bad will fails. The teacher may infect the will of the student by combining his love for the truth and his love for the student. If the teacher testifies to his membership in the fellowship of truth and at the same time keeps his membership in the play community which he has formed with the student, his testimony may take the boy up into the serious fellowship.

Since this is the core of the dialogue, it is worth the trouble to consult our present-day translations. We find that Leckie is uneasy when he is confronted with the bold sentence: *Tantum cuique panditur quantum capere propter propriam sive malam sive bonam voluntatem potest.* Leckie translates this: There is revealed to each one as much as he can apprehend through his will according as it is more perfect or less perfect. Augustine says, however, much more bluntly that the truth is spread as a linen or a rug, on our good will and cannot be spread if it would have to be laid upon a wicked will. We are so unaccustomed to the harsh statement that a student's will might be wicked and that only on a good will the linen of truth may be spread, that it is quite understandable to find our text mitigated in the translation. This character of the modern mind is brought out even more sharply in the translation by Tourscher: "It is opened out so far to each one as each one is capable to grasp by reason of a good or a bad habit of life." I doubt if the term Habit of Life, arouses in the modern reader the full sound of Augustine's word which sums up all our habits of life into "will." Habit of Life, it seems to me, is used by us too much in the sense of specific habits. Augustine calls a spade a spade.

To call wicked or evil will merely "less perfect," or the central direction of man's decision a habit of life, conceals the anti-logical character of the educational situation. The man of good will is the man who is open to the two forces: faith in the teacher, and love of the truth, without which learning cannot proceed. Reciprocally, the teacher must have faith in the truth, and love for his student. The dualism in the student is echoed by the dualism in the teacher. but not mechanically. There where the student has good will, the teacher employs faith. There where the teacher is bound by his interest in the student, the student is bound by his faith in the good will of the teacher.

THE CORRESPONDENCE OF HUMAN BEINGS

The second part of the De Magistro dethrones the teacher from his Lordship over logical truth. The great Guru in India, the heads of the schools in antiquity[,] were fountainheads of truth. Augustine insists on a triangular relation. God who is love and truth both, instills love in the teacher, truth in the student.

The modern reader will say: "Well, we know this. We no longer exalt the great teacher. We consider the teacher just one facility like any other." The student is admired by our progressives who tell him to be creative. Behind the child, the teachers disappear today as hired men, as the impersonal tools of the child's growth.

However, if Augustine's analysis is right, the modern attitude although topsyturvy compared with antiquity, is just as deficient as the pagan. Neither the child nor the adult carry the process responsibly. They can carry it only *correspondingly*. And their correspondence goes on in a medium common to both. Neither has the teacher a private claim to the truth which he has either heard or discovered nor does the child discover the world all by himself. When people think of a human relation as a purely dual relation, husband and wife, capital and labor, teacher and students, it nearly always seems to happen that the dualism soon is reduced by one faction to one half of the two, and by another faction to the other half of the pair. Labor says: I am everything, and we have communism, Capital says: I am everything, and we have exploitation. The husband says, I am everything, and we have the autocrat at the breakfast table. The wife says: I am everything, and we have—but I shall not say what. Now, in education, after giving nearly everything to the Guru, the teacher, we now hear people declaim about the learning genius of the child. In our age of the masses, the leader hides behind the masses which he leads, the teacher hides behind the sucklings whom he indoctrinates. Another fiction. This time, the truth is as much distorted as it was before.

May I suggest that all over our social world, any dualism runs the risk to be reduced to a monism when and as long as it is not interpreted as a trialism? Therefore it is of the essence that we understand the trialism as advocated by Augustine. Before I am "labor" or a "capitalist," I am a man. Before a man acts as teacher or as student, he is a human being. But what is a human being? How does the human being assert himself after I am disguised as a teacher, a husband, a capitalist?

The human being, not the teacher, is *bound*. As a capitalist, I "can" exploit or I "could" exploit labor; as a human being, I cannot. As a teacher, I can argue ad infinitum and sell my brand of truth like the sophists of all times, for big money. As a human being, I cannot. An exploiter, a communist, a reform school child, a tyrant, may deny this "I cannot"; they may shout: "In the devil's name, why can I not do as I have power to do?" Yes, why not? They all can overplay their social role, and we see them abuse it often. But is it not strange that the abuses do not range much farther? As a boy, I always pestered my father who had been to Russia and reported on the bribes and corruption under the Czar; with the one question: How can a country live in this way? How does one know that the bribe buys the goods? Why don't people accept the bribe and then simply refuse to make their promise good? I must have asked the question a hundred times. And my father always replied: It costs you from 15 to 25% of the sum under litigation; but at this expense, you are perfectly sure of the outcome. The abuse is in itself limited and restricted to this margin.

Now I understand what I failed to understand then. Even the corrupt judge, it seems—and he is I suppose the worst social weed of all—is bound by one little claim which he makes himself. He wishes to be called a human being. Even Richard III while he has resolved to become a monster, expects to be loved, to be called a human being by some woman. This terrible dependence of man on being called man, is the whole fence which prevents him from going mad with conceit or crime. As long as I pride myself on being a human being, I make two claims which are extremely difficult to push and to put over. One is that I have being, that I am real, and the other that I really am a human being. These two claims are just as bold as a claim to a gold mine, and as difficult to protect. Incessantly, others brush me aside as having no real importance, and that is, no being. And all the gossip in town, at one time or another, makes inroads on my claim to being human.

There exists an algebraic equation of a severity as 2 and 2 equals 4, whenever a man claims to bear a name. I call myself A; then I want to be called A, by others. Speech is a severe bondage. It is based on the golden rule that the name which I use shall be applied by others. When I say A, I start a mathematical operation in my community. I set out for an algebraic equation, holding on to my name A and the operation is going on until either the community has come round to my nomenclature and then: the equation reads: my A equals your A. Or, I may abandon my claim, and be satisfied with the name B or C conceded me by the rest of the world.

Now, I may abandon all particular names: American, Christian, teacher, lieutenant, and yet survive. But I cannot survive the loss of my two titles as "being" and as being human. If I lose my claim to the second, I am proscribed and treated as an outcast. If I lose my claim to the first, I am put in a lunatic asylum, as hopelessly unreal. So, any human being, to his ending day, holds out these two claims: Treat me as being real and as being human, and waits for the social algebra which bears him out. All specific social functions are mere surface roles compared to this underlying lasting role. This role consists of a correspondence between my names for myself and society's names for me. This correspondence binds us. Without it, we lose our being and our humanity. Most moderns take this correspondence so much for granted that Mr. Hitler was needed to prove to them that it was a perpetual miracle that this correspondence should make itself heard and felt. John Dewey, born In 1859, in the year of Darwin's book on the survival of the fittest, is so completely naive about the operations which in this year of the Lord, surrounded the birth of John Dewey, gave him his name, his schooling, his career, his freedom, and his reputation all over the world, that when we read his books on education, the humanity of teacher and pupil and their reality are taken for granted. He only wants to see them grow, and act intelligently. But grow into what: into chauffeurs who are so efficient at 100 miles an hour that they break all speed laws? into women who decline to have children because it does harm to their slimness? Foxes are intelligent, and weeds grow tall.

Nowhere in modern education a word is said about the roles which precede social action and intelligence and growth. The roles of being real and of being human, as a claim and a response, as a hope of the society, and an acceptance by us, as a name bestowed on us, and an equation of self-consciousness and social reputation. Because it all was quite safely assumed to be taken care of, in 1859 when John Dewey was born, and the struggle for survival was proclaimed.

But a human being does not struggle for survival; man goes to war. This is the very opposite of the struggle for survival. We struggle for other things than for our own survival. Why? because we hold on to the phantastic claim that we are real, alive or dead and that we are in a conversation in which we make claims or give answer to claims made on us. I know of course that the survival of our social group is today identified with the Darwin theory. But this is not true either. However, this is not the place to prove the fact that a man who goes to war may fight and die without this hope. We may be content with the obvious. A human being is not primarily interested in his own survival. No marriage, no childbed, no war, no religious persecution, no ordeal, no, not one of all these events, could take place ever, if man were primarily interested in his enlightened self-interest. Growth and intelligence do not suffice to direct our lives. Both are too self-centered. No man has ever lived by them, except the victims of pragmatic education. But we do live by the great human bondage which precedes any division of labor in society, and which stirs us into action and suffering and adventure and risk, all our life. This correspondence is like an unending conversation which is carried on with us. Elsewhere, I have shown that we do not start this conversation ourselves; the first thing we know about it, is a claim made on us.[8] We are called long before we call back. On the other hand, since this conversation keeps us alive, we are forever curious about the next answer, in this correspondence with the universe. It makes all of us thirst for some witness outside our transitory social function. Teacher or student wish to correspond to somebody outside the classroom because they wish to insure themselves against the loss of their human reality during the hour. The correspondence must get them outside their "roles."

It is of great historical interest to see Augustine unfold this primary relation of the man in the teacher and the man in the student to a third, corresponding voice. As long as either the teacher or the student think too highly of their own role in the process of conversing, they will say: "I teach," "I learn." These two expressions show a lack of correspondence. The medium inside of which the alleged two "Egos" find themselves is not considered. And yet, this medium of a common atmosphere is the astounding and tremendous fact preceding

8. *Angewandte Seelenkunde* (1924); "Modern Man's disintegration and the Egyptian Ka" (1939) [Available in *I Am an Impure Thinker* (1970, 2013)].

their own activities. "Atmosphere" is one of these wonderful academic avoidances of religious taboos. "Atmosphere" stands for common spirit, for what people breathe together, in as well as out. Atmosphere seems to be a natural fact; but since the term is nothing but a translation of "Spirit" it now to us becomes transparent as a social fact. The two, teacher and pupil, already form a "we" before they split into two Egos. Their possibility of conversing at all is conditioned by this common spirit which makes them meet with pencils instead of with shot guns. Hence the two Egos must be made to perceive this common basis, background, condition of one spirit.

He quotes from the prophet Isaiah the very word from which later Anselm of Canterbury took his "I believe so that I may understand" and which reads: "Unless you have given credit, you shall not understand." Augustine says that the student must first believe the teacher—modern theory notwithstanding—and from there go on to come in touch with the truth directly. We begin rightly by trusting our elders; in as far as they love us; they deserve our trust. Love is a claim to being trusted. But we must go on from there because God is not Love alone. He also is Truth and he asks us to meet him as truth as much as before we may have met him as Love. As truth we shall not meet him through other people's glasses.

All our qualities of a human being must be brought into play one after another. The teacher should not overtax his love, the student not overdo his faith. They must admit their greater partner, God, to their relation. Then, teaching is regenerated and converted and "rightly" treated. In teaching and learning, both partners undergo a process of reciprocal nature. We are cleansed of our distemporary limitations[,] the teacher by sacrificing to the future, the student by sacrificing to the past. Then, they have remained human, despite the moral risks of childishness and austerity implied in teaching.

THE BIOGRAPHICAL PLACE OF DE MAGISTRO

Let us stop here and raise once more the question: What does this dialogue achieve in the personal life of the two people involved in it?

A great teacher of the world, Professor of Rhetorics in the Roman Who Is Who? is speaking about rhetorics to his natural son. Adeodatus, at seventeen, is bright and mature. He is a real student besides being a son. This means that twice as much is put on this boy's shoulders than on the average boy who has to deal in the crisis of his puberty with a teacher here and a father there. Adeodatus is his father's student for years now. And this is not all. This same father and teacher has become a moral hero. He has dragged his son from one excitement to the next by taking him through the phases of his conversion to Christianity. Adeodatus went to baptism with his father. Where a normal child labours under one pressure, Adeodatus labours under three. Physical father, intellectual teacher, and moral hero, are present in one and the same person. It is true, the father had been baptised, and the son had been baptised. However, the relationship father-son was not baptised, so far. Augustine now was a Christian; Adeodatus was a Christian. Their fatherhood and sonhood were as before. Usually, a godfather and godmother take care of a child against the bodily parents, In our case, this was out of the question. Adeodatus was far too old, and had lived with his father all his life. The baptism happened far too late in life to protect Adeodatus against his father's spiritual despotism.

And here was Augustine, only 35 years of age, and his boy 17, both in the stage of fighting still. All the odds are against Adeodatus.

In this dilemma, Augustine himself serves the sacrament of spiritual emancipation to his son. And this is achieved by the dialogue. The dialogue ends on a tone which is unusual, personal, biographical. "I am not your real father, I am not your rabbi (= teacher), I am not your master and hero," these verses from St. Matthew XXIII become so eloquent in the mouth of Augustine. And he felt it; for in his Retractations, he sums up the whole dialogue after this quotation from the Bible!

He first shoots his summersaults together with Adeodatus in his respectable fields of grammar and rhetorics. He pokes fun at his authority as a grammarian and rhetorician. And then, he steps down or up to his real and serious role as loving admonisher. Thereby he completes his boy's spiritual emancipation. The history of the world hardly contains another case in which the words of the New Testament, these three verses 8, 9, and 19 against fathers, teachers, and bosses, resound with more meaning, more jubilance, more vigour than in our dialogue where they are meant to save the soul of Adeodatus. I do not know of any other case where a son was going to have his spiritual liberty sponsored and warranted by so imposing, so violent, so colossal a father. Would you or I have liked to be the son of Aurelius Augustinus?

Alas, his later students had no easy task. The dire need for De Magistro is proved by the besetting sin that we find in Augustine's own classroom at work. His disciple Eraclius preached in the presence of his teacher and the whole class one day and immediately went off in the same mood which Augustinus so violently refuted in De Magistro. Eraclius says: *Quidquid enim tibi in nostro sermone placuerit, agnosce quia tuum est; quidquid autem displicuerit, ignosce, quia meum est.*[9] This is the language in which our Saint might speak of God and himself, but which [he?] repudiated between mortals. The father did better than the bishop. This depressing example of adulation shows what the style of life still was and how very practical Augustine's considerations were. The contrast between Adeodatus and Eraclius may be taken as a test for the vital character of De Magistro, and our right of interpreting it as such. Augustine's own promise to give the positive doctrines later, is also a valuable testimony in this direction since it proves the programmatic character which he ascribed to our text.

To sum up, the sacrament of baptism of 387, is supplemented in 389 by the sacrament of spiritual emancipation. It has often been said that Augustine is personal. The whole history of the world, to him is his autobiography. The De Magistro is truly Augustinian. When we lift it to the level of a sacrament that purifies his last natural and pre-Christian loyalty, its form and content both are perfect. All other interpretations are at a loss to explain parts of the whole satisfactorily. When looked at biographically, the dialogue says: the Christian democracy is re-established; Teacher and Student move on one level of spiritual equality.

One cannot speak highly enough of the scientific potentialities eradiating from De Magistro. Many pre-Christian, pre-Augustinian fallacies about teaching linger in our classrooms. The greatest fallacy seems to me the most widely spread, namely that to teach logic means to be logical, or to teach science means to be scientific. This is simply not true, and we must be completely illogical, unscientific and irrational when we want to teach. For teaching is not indexed in the department of logic or science, it comes under the department of biography,

---

9. Migne 39, 1717 ff. "Anything you like in my sermon, recognize as being yours. Anything which you dislike, forgive as being mine."

and politics. As Augustine exclaims in the tenth book of the Confessions: "People must be connected by the bond of charity before they can listen and speak to each other with profit. "*Indicabo me talibus*" (Then I can show myself to them). Or as his disciple Eraclius said in the bad sermon with this one grain of gold: "What we see in him, is ours when we are in love with him." Teaching is charity, not thought; it comes nearer to the actus purus of charity than most human activities which are tainted by the will.

The difficulty of modern psychology seems to me the constant confusion between will and love. Psychology believes in the wrong pagan triad: will, reason, feelings, and love must be squeezed in as a kind of will which it is not. Love and will have as little to do with each other as a wedding ring with a gun. Will turns against external things, love is the creator of one body. How, then, can the oneness between teacher and student be explained in terms of will and reason? They form, from charity, *a body of time*; they are incorporated into an organism of time. A very practical consequence must be drawn from this distinction between will and love in regard to education. The pre-Christian world which is always around us, exalted the teacher into something of a hero or mountain of authority. The world of today does the opposite: teachers are discarded in favour of the student's self. We are told that the student makes all the discoveries himself. And the progress of education shall lead us into a time where the children need no teaching. Poor children. They will be cheated out of the body of unity in which old and young, teacher and student, become one. Both enter into one hour of forgetting time and space, by playing and thinking together, and therefore are released from fear. The hour from eleven to twelve in the classroom in a course of logic is a battlefield of reality, is a full present. The teacher is not teaching in the name of his science as Thomas Aquinas thought; he is not teaching in the name of a board of education or of the State as most people think today. Teaching has not any authority outside its own realm of charity and faith by which it establishes the fellowship between an older and a younger specimen of the human race. Teaching is the model social situation because it gains time. The contribution of the teacher's interest in the student, the student's faith in the teachers creates the time gain underlying society. Any sociology, that omits to put teaching in its centre, is unreal. That is why we have so many unreal sociologies.

They do not see that the gaining of time is man's political problem. Of this later. Let us first do full justice to Augustine himself.

The De Magistro must make up for a tremendous danger of Augustine's doctrines. To him who saw everything as biography, everything as transition and change in the human life, the soul is in every moment in danger of being nothing but passing.[10] The educational situation as I shall show in a moment is the antidote against too much temporality, too much transition and rush in our inner life. How can we avoid to overtax our poor soul by too much change? St. Augustine is anxious to put humanity in its place between the divinity and the world of matter. Change, history, progress is inherent to man; God is in eternity; matter is in space. Augustine literally says that time is the special property and qualification of man. You easily see how dangerous such a doctrine may be for the individual. Mere change is so fatiguing, so exasperating, because it makes you lonely time and again, from one of our ages to the next. Although growing in wisdom, man's growth must be balanced by achievement. This is done by the educational situation, between human beings. The experience of an old and the growth of a young person are welded in an hour of communication. In this hour, the partners are lifted beyond their individual age. They now represent two different ages, at least, in one "body of time." Together, they represent different tenses in the grammar of society or, with a favourite term of Augustinus, two different verses in the dramatic song of creation. The teacher and the student do not and cannot think the same things in this hour of communication. It would be blasphemy for a teacher to identify his thought with the student's thought. The itineraries of their minds are personal and must differ. But because this difference is survived and overcome, because the partners in the dialogue give each other three times, one to express experience, another time to grow, and a third time to communicate, they represent the model opportunity for man to have peace. By giving each other time, we communicate and become brothers; peace is nothing else but a state of society in which we are able and willing to

10. "*Quot optas gradus aetatis tot simul optas et mortes aetatum. Non* sunt *ergo is-tae. . . . Aetates labuntur, fluunt.*" *Enarratio* in *Psalmos* 127 Migne IV, 1686. [As many stages of life as you wish, so many wishes and deaths of stages at the same time. So they are not fixed. Ages slip and flow]. This is quite unheard of in the pagan world where the various ages of man were considered as individual blocks. Especially in India, each age formed an entity.

give each other time. In war, in the struggle for life, in the jungle, there is no time. When fellowship joins men of different ages, the times cease to be out of joint.

As an epilogue, or as a summary, I would like to look for a last time into the text. In chapter 14, we read that people are apt to overlook the time element in teaching. We perceive so quickly, it could seem that the teacher does what in fact the lapse of time does for the student. Augustinus says; Mostly (*plerumque*) no time passes between the teacher's exposition and the listener's grasp. Although this occurs perhaps in most cases, the fact that it does not happen always, is sufficient proof that it is a fortuitous coincidence. And the key to the educational process is furnished by the minority of cases in which time passes (*mora interponitur*) between the teacher's words and the student's grasp. This interval is precious for our understanding, and it may be given a special name; Richard Cabot for instance called it incubation.[11] Here we have a point which I recommend to over-accentuate in the future. This period of incubation is at the heart of education. Augustine allows for incubation. Our summer vacations allowed for incubation. It is barbarism to abolish them. To deal with time, between human beings, requires not less than all the three cardinal virtues. Faith is indispensable on the side of the Student before he can understand. Love is required on the side of the Teacher who must take an interest in the growth within the Student. And both must hope that their contributions meet in the opportunity to communicate. The reality of teaching is in need of all three qualities and of the three times. "The body of the time," to use the Shakespearian phrasing, contains past, future, and present in order to attain reality. Left to themselves, these times are abstractions. Incarnation is due to the possibility of communication. And Augustine's remark on incubation shows as strongly as his pet phrase: *Nisi credideritis, non intelligetis* [Unless you believe, you will not understand], and his combining love (charity) with truth, that all the elements of the process are keenly observed by him.

And his own book is the best illustration of his program. De Magistro is the full incarnation of two people in their biographical conflict and harmony. It is easy to define the beauty of this piece. A great man and an adolescent play together. In doing so, they eventually forget their earthly station as father and son, *magister* and *discipulus*, hero

11. *Ed.* Richard Cabot, MD, 1868–1939. See p. 78, above.

and follower, and go beyond their accidental roles. They move before us like two verses in one song of praise. And with an Augustinian notion, we see the beauty of temporal vicissitude, and see the orbits of their times associated to the song of the universe.[12]

FORMER EVALUATIONS

Our result is rather unexpected. At least, it does not coincide with the evaluation put on De Magistro by either one of the three groups that have commented on it. It is only fair to hear how De Magistro has been interpreted in the Middle Ages, in the Renaissance, and today. The extreme character of the three evaluations may well amaze us.

To begin with our own times, we may say that the De Magistro is remarkably popular. Mr. [Étienne] Gilson [d. 1978] gives it a number of pages in his study of Augustine. Twenty years ago. Father Tourscher published the Latin text; in 1924, he printed an appealing translation. Finally, in 1938, there was published a new edition by a friend of Mr. Scott Buchanan, George Leckie, which I must mention despite the shocking fact that Leckie does not mention Tourscher.[13] I must mention him because his long and very solemn introduction is the best illustration of what people in our days think that they can get out of De Magistro. Leckie's thirty eight pages of introduction deal with cognition, the liberal arts, especially grammar. The boy Adeodatus to whom Augustine is talking, the situation in which father and son were in in 389, after leaving their academic friends in Italy, are not mentioned. The doctrine of the book is investigated because Leckie believes that the Greek trivium, Grammar, Rhetorics, Dialectics, still offers ultimate truth to us, at least in the purified form in which Augustine presents them. Science, intellectual virtue, nor [not?] moral energies, emanates from De Magistro, for this school of thought.

Now, let us look back into the Middle Ages, to the Augustinian Bonaventura. His interpretation is condensed in a picture. You probably all are familiar with Fra Angelico's painting of the scene which might be called Bonaventura's commentary of De Magistro. Bonaventura who wrote the famous "Itinerary of the Mind to God"

12. *de vera religione* 23; *de musica* VI, 29.

13. *Ed.* Scott Buchanan (1895–1968), American philosopher and educator, known for promoting the Great Books.

in the Augustinian tradition, received the call of St. Thomas Aquinas. St. Thomas when entering his colleague's cell was surprised to find it devoid of bookshelves along the walls. "Where is your library?" he seemed to ask. Bonaventura withdrew a discrete curtain: a crucifix hanging from the wall, was his library. Christ was the Master of this great soul. Not just the teachings of the living Jesus as found in the Scriptures, to be sure; but the inner Cross and the inner Christ on the Cross were his books. The last words of our dialogue constituted the centre of the book for its medieval readers, not the trivial chapters on the trivium. Their earthly teaching was left behind much more definitely than in Augustine himself.

But it would be too simple, to see a dualism only; Bonaventura driving too fast on to the Christian goal, Leckie and the modern logicians getting stuck on the pagan road of the dialogue. For, we have a third tradition, that of humanism. In 1527, the Prince of the Humanists, Erasmus of Rotterdam, commented on Augustine's De Magistro. And in his few remarks, he gives the quintescence of humanistic criticism against Holy Writ as it has been applied ever since. He makes two points. 1. A few, plain truths of philosophy and theology (mark that philosophy has precedence) are obscured and frustrated by Augustine's skill in saying nothing in many words. The low scientific standard of his days led to this vicious performance. 2. The content of the dialogue may be reduced to the Platonic truth of the Logos, as the universal reason of all men. This Platonic notion has been quoted by St. John and was rhetorically expounded by Augustine. To this, Erasmus adds the maxim of all reductionists: This dependance should be carefully kept in mind by all readers of the Fathers; we cannot understand the Fathers without investigating from which philosophy they got their ideas.

In short, Erasmus says: What is good in Magistro, is Plato; and the form which is bad, is the only property of Augustinus. I was surprised to find as early as 1527 the same scathing method of the source-hunters that has dissolved in dust Homer and the Bible, the Nibelungen [,] and only by a narrow margin has missed out with Shakespeare. The Erasmus of every age reduces a text to its alleged sources; the text so reduced appears as a pure and poor contamination and loses all value. Well we shall have to face this reductio ad Platonem too.

Is Bonaventura right in forgetting the human relations of the learning soul completely, putting her behind a curtain with her one Master in heaven? Is Leckie right that it is the best basic doctrine for a renewal of the ancient world's ways of grammatical, logical and dialectical teaching? Is Erasmus right that the nucleus is Platonic, and that Augustine puffs this nucleus up rhetorically?

If any of these three judgments were right, I should not care for the booklet. However, they all treated the De Magistro as though it was written by Rhetor or Bishop. Therefore, they could not see the act of jettisoning the play-section by which act the book became biographical. True biographical acts have objective value. *Biography is at the core of sociology.* That is the masterful doctrine of "De Magistro." For all biographical events correspond. Our lives are reciprocal.

Undoubtedly, then, we stress an aspect completely neglected by others, and we neglect the aspects stressed by them. Yet, we may hope to justify our view if we can do justice to theirs. And indeed, these judgments were quite justified when we consider the central interest of the writers. . . . Bonaventura expected to meet the saintly Bishop of Hippo. Erasmus expected an imitator of Plato. Leckie thirsted for some solid foundation for teaching the elements of the trivium. They all concentrated on that element in the dialogue which represents their expectation.

After all, we did likewise. We concentrated on the biographical situation of Augustine and Adeodatus—in a vacuum between academic world and holy church. But we feel that we could do justice to all the parts of the dialogue; we did not have to be choosy. In the first half, the two interlocutors were distemporaries, one old, one young. In the second half, they lived in the presence of God, as his children. And in the light of eternity, their temporal differences had disappeared. The transformation of the two, from part one to part two, was the topic that put all the interpretations together.

The De Magistro—and I think, the variety of interpretations confirms my thesis—makes biographical reciprocity—an event in time—the core of education, of social life. We who are submerged by an economic, naturalistic, speechless, sociology in which education forms an annex to the "facts"—may take heart that a legitimate science of society has a sound basis and a great tradition. Where a man transcends his own time, there does he enter society. All societies create presents. The highest aim is to create the greatest, most comprehensive present. But the frail present created between Adeodatus and his father Augustinus contains all the elements which go with the most grandiose scheme of social organization. Here is the living cell[,] and a society which intends to live will consist of living cells or not at all.

THE CREATION OF A BODY OF TIME

The "De Magistro" is a dialogue in which something happens to the type called "a dialogue" itself. In the pagan dialogue somebody taught somebody else, proved him wrong, or proved, perhaps, that both interlocutors were ignorant.

When the ancient dialogue tried to become positive, it sloughed off its dialogical character. The late Platonic dialogues no longer were dialogues, but dissertations.

St. Augustine put this old form before us and employed it first as a playform of the human mind. Plato, too, used to play before he came to the point. But that which would be the content of the serious part in a Platonic dialogue, like Gorgias or Cratylos, is now the play part! Thereby, Augustine makes room for a *third* part (= his second) in which the dialogue itself is taken out of the hands of the two people who conduct it. The man of our era is in a position to know of the backstage from where the human drama is directed.

In our era, human speech has changed its character. In an inspired conversation, all the interlocutors may change their opinions during the conversation. The spirit moves freely. At the end, the one and the other may have changed roles and convictions, both. The words spoken are not to be put over by one, and understood by the other. The partners acknowledge a third power which does the moving of their minds and which allows them the complete freedom from their initial role or principles because their hearts are united.

This freedom is especially difficult for a teacher. Since in his case, the onesidedness of the direction of the current is so much in evidence: he seems to know; the student does not know.

How, then, is it possible to say that in a lesson, the two partners both unite in a third unifying element and are both equally changed? How is "teaching" truly reciprocal?

If the process was merely the exercise of our rational faculties, no reciprocity would be obtainable. The teacher would be a faucet turned on by a more or less eager or fastidious child to sip some bit of information. If teaching were information, the telling of facts, then the teacher would be a paid facility. And as a facility, teachers have been labelled by modern speakers on education quite regularly. If this were true, teachers would be the most exploited class of society, proletarians who should fight for losing their chains as bored and abused "proletarians" sucked dry by impertinent brutes.

The modern theory of education, with a bland front towards the parents, alumni, pupils, flatters their demands; and from fear of disquieting these customers, is silent about the moral status of a teacher. John Dewey actually allows the teacher to be merely a wage-earner. In his fundamentals of education, the teacher does not appear at all as a human being. He is a slot machine. His lubrication may come from heaven or from good pay; but it is not made the deepest riddle of the whole process. But why should anybody teach? Why does John Dewey write fanatically and inexhaustibly on education? Which passion drives him on? Is it a hidden, unscrutinized fever that makes him do all that he has done and does? Or is it a legitimate social energy and as accessible to investigation as the needs of the students?

If the teacher is not a real "liver" inside the educational field of force, if teaching means nothing in his life, then teaching must go out just as other forms of human servitude. A transcription over one big radio perhaps could replace it. *If teaching is work, let us have teaching machines!*

Everybody knows that all these assumptions are futile. Teaching is an integral part of any human being's true life. How is this possible if the teacher knows all the contents of his teaching before he enters the classroom, and if teaching is a rational process? Teaching would be mere duplication of my reasoning. Repetition, not life would be the whole process. Well it would be impossible if teaching were a rational process. But it is not. When John Dewey writes a book, he does so for utterly irrational reasons, for joy, pity, exuberance, sympathy, aggressiveness, hope, fear, for instance. And he does it by utterly irrational tools: patience, industry, justice, persuasiveness, learning, etc. Any teaching, when we forget emoluments, social rating, traditions of official schools and salaried teachers and all the things which make teaching a business, any teaching is based on three elements which place the people A and B in a time relation.

Let us now study this time relation as soberly as possible. A must be "older" than B with regard to the subject matter to be taught. He must have been involved in the matter before the lesson starts or he would not be the teacher. B is supposed either not to have been involved in the matter before at all or at least less than A. This makes B younger. Young and old are here clearly definitions of a relation to the theme of conversation. They have no foundation in physical age necessarily. The process of teaching forces us to consider "old" and "young" as relations of members of society to certain social experiences. *Old and young, are not biological facts; they are social facts.*

This is quite new and quite important. The ambiguity of "old" and "young" has concealed this social aspect of the terms too often. Of course, now after having defined our terms, we could use "teacher" and "student" again, instead, of old and young. However, these two terms are overlaid with prejudices at this juncture; hence we better stick to our social usage of old and young somewhat longer.

What do we gain by doing so? Old and young stress a time relation. Man grows old by experience. He becomes saturated with "process" which enters him, and in this process, he is consumed and finally dies. The old are nearer death; the young nearer birth. Not because the old will not survive perhaps many of the young but because he is more informed and formed and moulded. To be old, we then may go on to say, means to be full of form. To be young, means to be less formed.

Now, "form" means dying. The most genuine life in us also is the most shapeless. To be young means to give the formative powers in us free rein. Formative powers will set to work only in plastic matter. The old person has abandoned a part of his plasticity. We are as old as we are definitely formed. Conclusion: a teacher renounces part of his own plasticity for the sake of teaching. For, when we teach we must try to represent old age in the face of younger ones. A teacher needs something statuesque against which the waves of the future, the young, can break. [Anne Morrow] Lindbergh's Wave of the Future meant exactly this: an unbroken youth of the Nazi type, merely young and untaught by the experiences of older mankind, running on in waves of sheer youthful mentality.[14]

14. *Ed.* Anne Morrow Lindbergh, wife of the famous aviator, was the author of *The*

Why should a teacher renounce youth for the sake of teaching? Is this not the most inhuman sacrifice? Teachers are made old by their students. Any student thinks that his teacher has his life behind him. The teacher-student situation conceals to the young the sufferings and battles and uncertainties of the older man. Every student looks into [i. e., sees in] the teacher a kind of certainty and stability which the latter may not have at all. The classroom gives him the appearance of firmness, stability, certainty at least with regard to his subject matter. I always found that my students considered me infinitely older even when I was perhaps younger than they. What, then, does a teacher get in reward for renouncing his plasticity?

His reward is that he determines the future beyond his own time. With his interest in his student, he is effective after his own limited time and enjoins his experience on the younger generation. He sows into the physically younger a seed begotten on his own field of life; he conquers new territory for this experience or truth. When we listen to the call of teaching we are pulled by our love of an afterlife after our own individual death. "*Ne ulla virtus pereat*," Let no energy be lost, is a general law of reality. Forces which do not know of their own death, might waste themselves. The force "Man" cannot do so without sinning against the law "*Ne ulla virtus pereat*," because it is conscious of its own end.

Man is forced to teach, to transmit his experiences in the form of sowing them into younger men because the law of the conservation of energy plus his foreknowledge of his own death combine to make him seek an outlet into the future beyond him. Man, in other words, wants to determine the future. One form of determining the future is teaching.

The element which forces men to teach is then the connection a man strives to have with the future beyond his own time. There is in man, then, a time-arc holding out towards a time which he himself will not enter. By this element, man reaches out into a second time beyond his own. Let us call this feeler not with the trite name "Love" but with the most abstract purely chronological term of *the forwardizing force*.

This forwardizing force is not to be thought |of| as mere expansion into the future. It is based on the assumption of a break between

---

Wave of the Future (1940), a defeatist and pessimistic work at a time of crisis.

my own time and the following time. The future is somebody else's time. In teaching, the relation between present and future stands revealed as the relation between my life time and the times after my own death has occurred. A definite break is posited between present and after-present, and my knowledge of this break produces in me the forwardizing energy called teaching by which part of my experience can be regenerated in somebody else.

The analysis of a teaching man clarifies the relation of our time sense to our death-consciousness. Man knows of time because he knows that his own time is limited. Hence, he is forced in every moment of his life to distinguish between his own lifetime and all times beyond or after this inevitable event. Man handles two times "all the time," so to speak. And he tries to transport as many particles of his life from the section of his own time to the section of time in general as he possibly can. The mechanic aspect of time as uninterrupted flow is not to be found in us. *We have a split time sense.* And the present and the future are separated by the grave. May be that this grave is not very tragic and does not even include our whole man. Between this present extant moment when I teach you the ABC, and your applying it, not much time has to elapse. I may very well live on beyond this lesson, and forget all about it. But still there yawns an abyss between the present of this lesson and the future although no complete physical death intervenes. The abyss simply means that my energy has found its outlet into a future beyond my own time, by telling you. The importance and beauty of the ABC made me wish that it should not be forgotten. I did something about it by testifying to its importance and by insisting on your taking it in. As soon as I have done so, I feel relieved, and I feel free to forget about this part of my experience. An experience successfully transmitted to others frees the transmitter from the burden under which he labors before he has gotten the transmission out of his system. And so, even in the most superficial form of teaching, there is a break between present and future. The forwardizing energy when it has left me, leaves me with a feeling of freedom I did not have before.

Man does not live in the present alone but, by merit of the forwardizing energy, he reaches a beyond-himself time. The teacher is forced to enter a relation to human beings whom he can teach because he must make this connection with a beyond-himself time. Once he has determined this beyond-himself time, he is relieved.

Now, the pupil, too, is not shut in into his own lifetime. He, too, holds out an arc of time into the times beyond him and seeks to make a connection there with other times. But as a student, I try to make this connection with the past; I backwardize, chronologically. I wish to experience preceding experience. If I would decline to learn, I would be a brute. Nature has not found the secret of teaching the young the new experiences of the old. The transmission of newly acquired faculties is the privilege of a small part of nature's chaos, especially of man. Man is he who can inherit faculties acquired by other members of the race.

The pupil, then, is not compelled to go beyond his death but he wants to get before his birth. Again, the term ["]birth["] covers a multitude of situations, as the term ["]grave["] did before with the teacher. I wish to learn how to ride. I must learn how others did ride before me, before the hour of my being born to the horse, so to speak, struck. It is the relations to specific experiences into which birth can be subdivided. As many varied experiences I undergo, as many births occur in my life. And as many times shall I try to learn the antecedents of this my new birth to this specific matter. I wish to get back behind my birth, into my so-called "background."

In other words, or to coin a purely chronological term, a young man who learns, penetrates into the before-himself time, by backwardizing. He holds out a feeler into the past. He is compelled by his consciousness of birth to go back of his birth. Before him, men lived already. Whether he likes them or disapproves of them, they have formed all the matters and objects and words and laws and habits and rituals which he may conform with or reform. His freedom depends on his getting back of these forms into the time when they were still in process. To learn means to go before the forms into their formative moment. Because then, the past and my background cease to be rigid determinants of my own form and habit. *In backwardizing, we re-enter the ranks of those who determined the past.* The parallel to the teacher, then, is quite literal. The man who teaches determines the future by his experiences. The man who learns determines the past instead of being merely determined by it. And he distinguishes therefore between past and present clearly. There is a break between his present and the past, a break caused by his "birth." To backwardize stresses the chronological aspect of the faith we have in the world as we find it, into its character as good, as a created order to be appropriated by us as heirs. To forwardize, stresses the chronological aspect of the hopes we entertain with regard to the future, to a time when we shall not be wasted and not have lived in vain, but form an integral part of reality, and remain inscribed into life as founders.

Now, how to get teacher and student together? One holds out his feeler into an "after-me" time; the other feels his way into a "before-me" time. In the hour in which they communicate, they build out of these two elements a common present.

The individual A, in his own time, plus his time-arc into an after-himself time, and the individual B, plus his time-arc into a before-himself time, step together on a platform created by their common effort but not existing outside this effort. During the hour during which pupil and teacher converse, time is forgotten in a very definite sense. What is said during this hour from eleven to noon is all simultaneous! What a teacher says at eleven twenty and eleven forty, does not belong to different times. How is this to be understood? Obviously, in physics, the moment 11.20 and the moment 11.40, are considered outside of each other. They are disconnected and they are separated by innumerable other moments between them. But this is true of physical, external time only. It is quite untrue of the classroom hour. During this hour, times are inter-twined which in one individual cannot be found. Alone, by myself, I cannot get before my birth or after my death. In the classroom or in any situation of teaching, I can. Here, the teacher in his impulse to reach the time shore which lies beyond his life (or beyond the liveliness of a certain period of his life with experiences which crave for succession), and the student in his impulse to reach the time shore before his own bursting into consciousness, are moving in opposite directions: one backwardizing because he asks for his first cause, his fatherland, his mother tongue, mother nature, his alma mater, his roots, his pedigree, the "evolution" of his universe and how all these innumerable labels for the "before-my-birth" impulse run. The boy and girl in us always ask: why? which is the birth-question. The backwardizing impulse of a person makes this person into a young person. To be young, is a relation to the past by which I try to unveil my mysterious antecedents. An unveiling tendency towards the world proves that this animal tries to become human by getting behind himself, by becoming its own author.

On the other hand, the quality of "old" means that the forwardizing impulse is active in a being. The senior's animality tries to become human by reaching the time shore in front of it. The shore beyond my lifetime carries as many labels as the "why"-front? Hence we have an accurate correspondence of old and young. The causal front of the junior who asks "why" and pierces the wall before his birth is not more numerously labelled than the "purpose" front of the old with its "whither?" This front contains all the "oughts" of the ethical code, all the anger of the sterile who are not in a harmless contact with future generations and who therefore get to the young only by slandering them. It includes all the reformers, revolutionaries, rebels, radicals, "-ists," endowers of gigantic plants with no soul in the buildings, and also the real parents and ancestors and legislators of the future as well. They all try to become their own apostles. We all are driven beyond ourselves by craving authorship before our time and embodiment after our own time.

Now whenever one man's "why" and another man's "whither" can be soldered together, something happens. They found *a body of time*. It's like a pipeline for the stream of consciousness, our psychologists might say. But I fear, the metaphor is not rich enough to make clear what really is attained when a "whither" and a "why" soul, an old and a young person fall in hope that they can help each other out.

Any Body of Time constitutes a fusion by which one's time in the form of future and another's time in the form of past are made accessible to each other by hope, faith, love. Without the mobilization of these three energies, the animal cannot become human, and the roving individual cannot ascend to the quality of reality, of being. It is the condition of your humanity, reader, that you read and write, listen and command, ask "why" and feel answerable for the "whither," that you contain the two elements of old and young in you. The three tenses of grammar: past, present, future, do not exist unless the three energies or potencies called faith, love, hope, have become activated and effective.

We have discovered the great fallacy of our own humanistic tradition. Humanism accepted the division into past, present, and future as a natural fact which seemed to be inherent in the world outside of man.

Humanism was mistaken. To divide time into past, future, present, is a creation of society. It is an expression for the "supertime" which comes into being when more than one generation are made co-existent with each other. Wherever young and old learn to coexist, a creation takes place which allows them to contribute their two time horizons to one pool, inside a common hope. Inside this body of time, that into which the young wishes to penetrate is called the past; that into which the old desires to advance is called the future. But both, past and future, are qualified in that they remain outside the real grasp of the desiring individual itself.

The historical or social creation mediates between this frustrated individual and the time shore which he is longing for, by binding him to another individual with the opposite "time-shore" complex.

In this binding process, the span over which the two oppositely facing individuals overlap, is the present. I have as much present as I contain meeting ground within myself between my great great grandchildren and my great great grandparents. If I can hold a meeting between Old Methusalem and the man of 5678 A.D., in my chest, I am representative of so much past and so much future that my present is extensive. *The present does not exist in nature. It is a gradual product of the three cardinal time-producing energies in society*, and it has to be reproduced incessantly. The present may be lost. And then, the world breaks apart into individuals who are neither young nor old but unteachable urchins and unimpressible martinets.

Since the great fallacy of the "scientific" era took ["]time["] for a natural we contemporaries of two world wars now have to use the term "supertime" for the processing of articulated time. Natural time is inarticulate. It is so fleeting that we do not bathe a second time in the same river. The supertime allows for the articulation of one part as past, another as future, and the common life in between as present. The supertime is the superman in that it builds two "time shores" before and after the flood of my own life. These shores are assured only if I care for predecessors and successors of "my" life and am willing to identify myself partly at least with ways of life which went before me and shall come after me.

Past and Future "are" not. They are a process of frantic waving backward and forward and enlightening our comrade in hope about the time shore from which we stem or towards which we are heading.

It is necessary to replace "past" i.e. the world of facts by some such word which expresses dynamic movement backward, and "future" similarly by a term of process. Science has preempted the two terms "fact" and "future" which come from the New Testament. This Christian origin of the two terms is overlaid by the intervention of science which inherited them from the Church. In the New Testament, the "future" was the time in coming, the time shore beyond your or my life. And the "fact" of which Jesus spoke when he said, "It is done," was the time "done" which from then on could form the background of every child born into our era. There now was one fact which formed the background back of every human being, a fact older than anybody's life who would come into the world after Jesus, a fact to which the sons of Adam and the daughters of Eve could look as naturally as they looked before to all the pedigrees of their tribes. They now were the brothers and sisters of one who had placed himself between all past and between all future, between all backwardizing impulses of the young and all forwardizing impulses of the old.

Antiquity did not know of any way of conquering time as I have shown elsewhere. Antiquity lived in mere cycles and ceaseless revolutions. Future and progress were unknown.

No wonder that those institutions of ours which we took over from the ancient world, try to ignore the social character of our time notions to this day. The academic world which is Greek in origin still cultivates a disdain for supertime, and for the energies which alone are able to produce it. Faith, Love, and Hope, are not considered worthy of scientific consideration. They are called irrational, unproven, non-existent, cobwebs of mystics. They are left to Sunday school teaching, by these humanists and scientists who in their timeless academic world look down upon the people who have to righten the times in war and peace and who support science, by their faith, love and hope.

Faith, hope, and love must have done a lot of special work before one course on logic may be announced at Wabash College. Love, faith and hope are the most real, most practical, most tested social energies. In fact, the virtues of which the world makes so much, justice, prudence, temperance, simply could not work unless there first is a society welded together of a number of generations. These generations must have created a cooperative present with common hopes, common past, common future before the pagan virtues can adorn the city. Supertime precedes the humanity of each individual. We become human by entering into a body of time.

We now are able to give these definitions:

A body of time is the product of social cooperation of at least two individuals who create a tripartition of time, a supertime which consists of a present between a past represented by one of the members of the team, and a future represented by the other. The present is held by both in a clasp of mutual trust.

When we speak of past, future, present, we always presuppose a more than individual or biological time. The supertime is based on the display of faculties in man which he owns in as far as he has a stake on time shores which lie in back and which lie in front of his own time.

Time is [at] a standstill as one elongated extant moment, in the classroom, when and while the real past, the real future, and the real present are contributed. These tenses are real when they are separated from each other by a clear break, through the recognition of an interruption between past, present, future. The interruption is realized as a birth, between present and its background in a past; it is realized as a death, between the present and the after-me future. Birth and Death may be restricted to the birth or death of one particular part of the man. But it remains essential that man realizes them as absolute lines which hinder him to cross them in the flesh. Before we become "real," we must have tasted birth and death. Reality is not to be had without first realizing our time shores.

I cannot get into my background, into the formative energies which moulded my plasticity, in my own flesh, but only by faith in other people. And I cannot get in front of my own lifetime, except by my love for other people. But I want to get back and forward. I and everybody else backwardizes and forwardizes with might and main and we recognize each other in this human trait. And our hopes that one may help the other, draw us together. And we do create a timeless present.

The educational process is only one specimen of this social creation of time at a standstill. It visibly connects only two generations, one old, one young. And that is very little. But it is enough for a scientific study of the "time cell" out of which the whole body politic of society is composed. The educational situation is the smallest atom of supertime. The actual time of man is experienced in four dimensions: as past, as future, as physical fleeting moment, and as the standstill present.

The past means a before-my-time background, to be conquered by faith. The future means an after-my-life-time to be conquered by love. The timeless present is based on the common hopes of distemporaries that by pooling their time sense, they might become contemporaries of one standstill-present. And how about the fleeting time of science, this strange dogmatic concept of time which is used in our astronomical reckoning? Fleeting time becomes observable on the outside of any group which is bound into one supertime. The Eastern Standard Time is the external time which a supertime society fastens on the objects which are incapable of entering a present. That which we cannot incorporate into society, our outside world, is measured by this artificial time which its objects can't realize.

A "timeless" present is unknown to man. What is called so, is a standstill present in which the past, before birth, the future after death, and the physical present, are made coexistent. The success of this blending of the three tenses consists in our freedom to treat an earlier moment as later and a later moment as earlier, and to treat large stretches of time as having all one and the same "time," of being contemporaneous. Whenever I can say, that the first moment and the last moment of an event, are actually indiscernible in their time character—for instance in a piece of music, or in a movie play, in which beginning and end obviously form one indivisible unit—I have attended the successful creation of the standstill present which is a blend of the three tenses, past, future, present, into one compound time. The success is based on the contribution which we make by bringing our time energies into play. Where we backwardize and have faith in the times before our birth, and forwardize by loving the times after our death, and have hopes in the present moment, we may create this standstill present by becoming representative of mankind in general. A representation of the whole including the before-myself and the after-myself, has the soothing character of bringing time to a standstill.

He who does love the times after his death by energetic for-wardizing is not rushed by his constant fear of death. He who does not fear the times before his birth, is not haunted by spectres and ghosts of his imagination.

But what do we see? Innumerable people seem to be haunted by the "before-themselves" and the "after-themselves" times. Psychoanalysis of the young and rejuvenation cures of the old are the two most advertised processes of our times. The standstill present between the generations must have broken down. Man is rushed and haunted, in even proportion. Could it not be that the misinterpretation of teaching and education has its share in this distortion of human times. In this case, the Augustinian dialogue might form the rallying point for a science of teaching and of society which includes the teacher's own life problems, and thereby makes education the nucleus of all social life.

As the coexistence of more than one moment of time and as the articulation of the many times of man are the foundation of education, so is this same coexistence the central fact between nations, races, classes, professions. All our activities end [and?] functions in society, entered society at some day in history in a powerful birth and eruption. And all which proved their mettle had to rejoin the standstill present beyond their own partial existence and scope. They all entered a wider tapestry of life in which their background as well as their after-their-lifetime future were reconciled to their peculiar time span. No way of life can pose as the first or the last way of life, not even "the classless society" or the master race who both tried exactly this!

In this sense, my Autobiography of Western Man showed the coexistence of many "times" which the great nations of the West undertook to represent, one after the other.[15] Each one of them broke into reality with an absolute claim of being the absolute and only form of being. Each one of them, in the course of an educational process was made a member, a relation of the family. We, in our days, are experiencing this process in the story of the Russian Revolution. Out of the only and absolute truth about economics, it has become one truth. Living at first to itself as though in its own times it was the only contemporary and as though everybody else was antiquated simply because his way of life had preceded Bolshevism, the Russians now

---

15. *Ed. Out of Revolution. Autobiography of Western Man* (1938).

join the Virginia reel of time, as one partner in the vast conversation of the whole race.

In many other monographs, on the polychrony of the people, on our economics, on the forms of government, on the shift in emphasis to various phases of our life cycle, the same fundamental law of time, of human time, was verified: Man can insist on one element of time, on his own time; but he has to become a correspondent in a body of time or his specific time will be wiped out. Times which wish to make a lasting contribution, have to develop the specific organs by which they can embrace their own predecessors and successors in time.

Each period, at first, is drunk with its own purposes. But each period is more than any one of its own purposes. It also is a tone in a symphony. Love, faith and hope, are the ears and voices by which the individual time can speak to all other times and be acclaimed by them as their brother. All the innumerable times receive their consecration from this acclamation and from their addressing themselves to all other times. Hence, the situation between one teacher and one student, is the paradigma of all social happening which comes into the world with the power to stay.

In the De Magistro, the creation of a common, standstill present is experienced. The ancient form of the Platonic dialogue is transformed into a biographical scene in the life of father and son. The son is emancipated. The dialogue is converted into an instrument of complete freedom.

Perhaps this explains why Augustine could not compose the little library for his son which was to follow. The De Magistro had done something beyond formal instruction. It had made epoch. *And an event of a strictly epoch making or biographical character cannot be repeated.*

We shall never know this because Adeodatus died and Augustine became a bishop. And as our terrible example from Augustine's own student has shown, inside the institutionalized church, teaching was deprived of this biographical character. The church in teaching the secrets of the creative life and the standstill present, teaches them in a non-creative and pre-christian manner. And she does so to this day, either in the Aristotelian forms of the middle ages or in the Platonic manner of the Liberal Arts college.

The Church as Magistra in as far as she transmits her secrets, is not yet converted. *Ecclesia Magistra est nondum Christiana.* [The Church is not yet Christian]. She teaches Christian things in pre-Christian, timeless style.

It was only in the interval between his Greek philosophy and his Christian theology, that Augustine for once, dropped the distinction between living and teaching, from love to his son. The dialogue is like a flash, holding out the promise of a life in which even education would not talk about life or report of life but be something in the life of the teacher.

There is a very simple criterion which shall show when this happy moment has come. When people today reform education—and show me the person who does not reform education—they discuss and plan and try to sell their new wares to parents or students directly as the case may be. They never take time out to become first of all new teachers themselves. The institution which would ask first how the salvation of the teachers can be achieved, and what they feel they must know and must transmit, and which would give infinite time to this question, would inherit the spirit of Aurelius Augustinus, not of the philosopher, not of the theologian, but of the man. Since we meet Augustine here as a human being, as the swimmer to the same time shores which we also have to reach in order to become human, the new science of time and supertime, is free to become a universal science.

From the frog's perspective, the eternal Body of Christ which Augustine entered as bishop, is visible only in the fragmentary form of the individual's limited contacts. I realize in my own lifetime only sketchy and small bodies of time through the precarious fellowships in which I am allowed to move. Hence, a science of time which expects to be recognized by us little frogs as dealing with facts, cannot begin in the sky of the most comprehensive experience of all mankind. It can appeal only to the minimum of social experience which every human being has because he has been called and wants to be called; because he wants to be real and wants to be human.

A science of time stands on the shoulders of Augustine but it cannot help being strictly secular.

On the other hand, this same science of time is as opposed to the method of natural science as science itself once was to the methods of the schoolmen in 1600. For the philosophy behind all science knows of fleeting, external and objective time only. It ignores the creation of supertime although it depends on its creation by society before any artificial Greenwich time or Eastern Standard Time can be established. The Republic of Science is occupied with nature, and that is, with the leftovers of the body politic, the objects which offer too much resistance to being incorporated into our body of time. These objects preferably are inanimate, are mere bodies of space. Hence the natural science has developed methods which apply to bodies in space only. Such science of objects is without any method to cope with the task of creating supertime. Science takes supertime of families, countries, nations, churches, schools, for granted, and matches them by its objective time for objects. The secular science of time is, then, neither philosophy in the sense of the sciences of the Renaissance and the Enlightenment nor theology in the sense of Thomism or Bonaventura. And this is the reason why we claimed at the beginning that the De Magistro and the science of time belonged to a third Body of Time, to our era in which we must rediscover the heat which begets supertime.

———————————

By speaking, we create something which did not exist before, and this something is society itself. The analysis of Augustine's dialogue enables us to see this process in operation in a small and isolated experiment. The experimental character of the group which comes into existence through this dialogue, is, however, a model situation for all human relations.

We speak in order to create a common space and a common time between people who by this creative action are transformed into beings and humans, or more briefly, into human beings. These common spaces and common times precede the secondary concepts of time and space used by the natural scientists.

The supertime and the superspace of which we here heard, must have been established before anything that is said by Newton or Einstein or Kant on time, makes any sense. Science uses its power to organize the things of nature outside[,] in a time and space continuum[,] as a delegated power. The power is delegated to science by society because it exists as a supertime and superspace through dialogue, correspondence, and conversation of all its members.

The physicists sit inside this supertime and superspace and look out from it into the world of objects. In order to measure these objects at all or even to observe them, they persistently borrow one fundamental creation from society which does not exist in nature. The unrest in modern science comes from this naive loan of a quality of supertime which mere observed time or time in nature does not possess.

[Pierre-Simon] Laplace [d. 1827], the author of the famous hypothesis on the formation of the universe, wrote one sentence which gives away the precarious and dependent state of science. He wrote: "We ought then to regard the present state of the universe as the effect of the anterior and the causation of the one which is to follow." The very word "ought" would suffice to the critic that social obligations here must have been pre-established. But I do not stress this now. But the relation of the tenses of supertime is here completely relied upon. Because in society the generations embrace each other beyond their individual birth and death, they receive the peace of a present between one generation's future and the other's past into themselves. Science then, for its external research, reverses the sequence and speaks of past, present, future, in this order. But this turning of the glove outside presupposes a glove woven for the human hand first. And there the order is Past, Future, Present because the knitting together of more than one individual's time spans alone makes the present come to exist.

Supertime is created by correspondence between human beings. This correspondence is operative when people converse. The analysis of supertime, for this reason, opens the path to an understanding [of] human speech. Speech was not made to think out loud. Speech was instrumental in the creation of a common space and a common time around people who individually have no time and no space, or for the modern, already socially sophisticated observer, have an isolated time and isolated space.

In speech, the homo sapiens of the animal kingdom can keep his unity through all his innumerable individuations of births and deaths. Through speech he is one man, the everlasting man, the human being. He is moved into a not objective and not natural common space and time.

This power of speech is not an appeal to man's rational or intellectual faculties only although it appeals to them too. But it appeals to the whole man. Speech is four times as rich as thought. And without this wealth of appeals it could not move man into superspace and supertime. Society is built by the energies which enable us to get outside our own short living time and living space and which make us to desire to melt into the world, be born into the future, enter the graves of the past, and reach our own innermost centre.

Speech, far from expressing a man's thought, enables him to think at all, as a representative of the One Man of all times and spaces.

If this is true as Augustine seems to suggest, then the study of language must reveal this truth to us. It must be possible to explain the structure of all language by exactly this one criterion: Does it create supertime and superspace? If so, our thesis is verified. If not, it is refuted.

We have examined the linguistic material. And we shall try to determine the necessary concepts of a universally applicable grammar.

Four

THE FUTURE WAY OF LIFE

Christmas 1942

I.

A young man of my acquaintance tried to become an ambulance driver in Egypt because he considered himself a Christian Pacifist. This plan failed; he was drafted. Whereupon he found that he was not a real Pacifist. But now, he went to the other extreme; he left the Episcopal Church. Not being able to live up to his own allegedly Christian standards, he decided to become a "pagan" all round, without trimming. It must have broken his heart, this absurd decision, because the daily service of prayer and praise used to fill him with joy.

Everybody knows that thousands of good boys are in a similar predicament, though perhaps not that extreme. Why not help them? But I find that most people are embarrassed. I see no reason for embarrassment.

Certainly, if our young soldiers now consider themselves as "fallen men," and confuse this fact with paganism, it is not their own fault. The Civilian tradition of thought during the last decades did little to encompass the character of the soldier. And so, noble energy goes to waste, in many souls.

The place of the soldier as against the thinker, the scientist, has to be illuminated and to be redefined. Since the millions are soldiers now, this mighty republic will lose its identity with its scientific and rational past if the connection between soldier and science is not found. The era of science must allot a reasonable place to the militant fighter. How can War and Peace be lived by one homogeneous and catholic people of thinkers and soldiers? Instead of our cutting them into two frustrated halves of brutes and brains, how can the human soul triumph over the separation of thinkers and soldiers?

Is it true that "the only answer seems to be to have a completely separated body of people to do the thinking and planning of which the average 'warrior' seems incapable"? This sentence is from a naval officer's letter, after thirteen months of service on a destroyer in the Pacific. And he goes on to say: "Somehow I don't seem to be able to reconcile the ideas of responsibility to the immediate job at hand and rather complete irresponsibility to the greater task of creating a new and somehow better world. . . . My greatest need at present is to talk seriously with someone outside the service who is concerned more with the future than with the present contingency of shooting down an enemy unit associated with a machine which must be utterly destroyed then *forgotten* while one concentrates on the means for eliminating the next unit."

And his letter continues: "The thing that has struck me as most dominant in the 'soldier' is the idea of 'forgetfulness'. This characteristic is most evident in our men . . ." (follows description of battle scenes).

"I know you can help me to be a better 'soldier'. If you don't maintain some sort of an ideological structure for us to gather up when we return I don't know what will happen—but I feel certain that you won't let us down."

Accordingly, I shall try to redefine the interaction of soldiers and thinkers. For this purpose, the current identification of war with paganism and of peace with modernity will have to be abandoned.

Our fighting boys are not pagans, and our conscientious objectors are not "more" progressive. In clarifying this confusion, a new definition of paganism was found by another friend of mine, a college senior. And his case may start us off on the discovery of the future way of life. Instead of being confused by a welter of experiences, he had been chilled by a dearth of reality. He started, then, at the opposite extreme, but ended up at the same solution: a way of life which would include peace and war, reason and faith, thinkers and soldiers.

II.

"Is not the whole Old Testament a collection of paganisms?" These words were said without much emphasis by the college senior. When I mentioned this feat of ingenuity to his teacher of English, he said, "Arthur probably did not know what the word paganism means." However, it is not quite so simple in this man's case. His great grand-father was College President, his grandfather headed a very famous Theological Seminary, his father is a missionary and the head of a big school. So, I do not think that it is with him as with most of the young who do not know who Abel or Moses or David were, let alone Ruth, or Isaac, or Absalom. No, he meant by paganism something bygone and insufficient. He did not use the word as praise. But he thought that he could sit in judgment over the "paganisms" of the Old Testament. He believed that he had inherited Christianity without the impact of the "pagan" Old Testament, and without a personal acceptance of the continuous order created by our faith. He did not know that anybody makes himself into a pagan when he mistakes the belief in the trinity for a hereditary property. He then becomes an "absentee owner" of the spirit, and such an attitude is as disastrous as the Irish landlord's absenteeism.

Most people I know who still care for the New Testament, poohpooh the terrible Jehovah as though he were another kind of God than the one revealed in the further quest. Their own sugaring of the deity, ending in pacifism, ethical culture, humanity, led these friends to an overconsumption of sweetness and an underconsumption of brain.

It would take a book to pursue the consequences of the "Sweet Jesus" heresy into each department of life. It certainly makes the flow of life into the future impossible. Half of our people, nine-tenths of the world, are at war. The God who is a God of War as well as of Peace will crush the worshippers of their own "wishful-thinking-about-God" easily and rightly. These "Marcionites" in the church of today stir up the persecution of the churches by those who are and have to be at war, dangerously.[1] Because these men are defeatists. Marcion had to condemn marriage just as much as wars. Property, children, family,

---

1. *Ed.* Marcion was a 2nd century c.e. church thinker who denied Christ's corpore-ality and human traits and condemned the Jehovah of the Old Testament.

and war go together; the pacifists must forego these goods or they are insincere.

But Arthur was sincere. No more sincere boy in the world. And he asked because he still felt that he might be wrong. What should I do?

If I show him, I said to myself, that the first verses of Genesis reveal the secret of the Trinity—Father and Spirit, Father and Son—he will not be impressed, I suppose. This generation will not be impressed by any theology whatsoever. (And the reader may breathe more freely from now on; he will be spared denominational squabbles.)

So, I turned not to the beginning, but to the end of the book, and I said: "You can read the whole Old Testament, or the first five books, or any of the prophets, or the first three verses. It says the same everywhere. But in your own language, it speaks in the last two verses of the last book of the last group in the whole. They are ascribed to the prophet Malachi, and he must have been as sociology-bitten as you are. For he expressed the secret of the Father-God, Spirit-Father, Father-Son, mystery, in sociological terms: "Each time," the prophet said, "when the hearts of the fathers and the hearts of the children are not turned to each other, the land is cursed."

The land is cursed today. The hearts of the parents and the hearts of the children do not entertain the same hopes and fears. For example, five years ago, a minister in New York City established a common living center for graduates from the great Universities. And when they gathered in 1939 before the war— these young businessmen and lawyers—the wave of the future had engulfed them all; they all were for a mild form of Fascism. And they certainly had no religious or political convictions in common with their parent generation. They had perhaps no very definite convictions, themselves. But, instinctively, deep down in their impulses, they expected regimentation and were ready to be regimented. And their soul was of no great concern to them. Now, this situation of the young is a biological fact just as is the fact that you have two legs; each generation has its predilections. The previous one had a social worker's tender conscience and a belief in freedom. And that again is a fact. These two facts are pre-religious, pre-christian, pre-historical. For no one generation can make history all by itself simply on the basis of its predilections.

III.

If one generation could be alone on this earth, and could forget about their fathers and their children, they would need no religion, no revelation of Father, Spirit and Son, Obviously, every one of these words connotes some process which is not identified with a single generation. The terms are clear. In the Divinity, Father and Son unfold the quality of being, by spreading it through two generations. And the Spirit, lest he be confused with the "wit" of the moment, is explicitly said to ascend from the interaction of two generations, the Father and the Son.

Neither the social workers nor the fascists have a future. For, the ethics of no particular generation can survive that generation. Ethics, after all, are mores. And mores, thank God, are transient, when they are good mores.

When I showed Arthur the sociological expression of our faith in Malachi, he immediately understood that neither paganism nor Christianity had to do with ethics, but with man's attitude to the products of his own brain. The pagan worships the products of his own genius. The Christian worships his maker's creative spirit. The pagan equates the life span of his own generation with the life span of that energy which allows him to think. And it is the pagan's obsession to believe that his mind is generated within him at his birth. But this is not so.

Man's life span goes from birth to death, and so, he usually thinks of his mind as being his lifetime companion, progressing also from birth to death. However, while this life stretches from the cradle to the grave, the life span of an inspiration reaches from the middle of one man's life to the middle of the life in the next generation. Thus, the difference between our physical and mental existence is expressed in the difference of their periods or rhythms. As carriers of physical life, we feel our life to be an unbroken sequence. As carriers of valid thinking, as scientists, rulers, writers, parents, experts, officers, we can't have peace if we try to imprison this thinking process within us or if we think of it as synchronized with our own biographical rhythm.

These ways of thought invade our physical existence in the middle of life, mould us, put us into our class, vocation, or office. And by our functioning in them and under "the rules of the game," we start younger men on the road to succeeding us in our social role. Shaped life attracts younger, more shapeless life always, because in nature all shapeliness commands reiteration. (The psychoanalysts call this the "compulsion to repeat," but it is the great economic law of the universe.) The young always try to inherit everything which there is to inherit from the past.

For instance, every boy or girl in this country learns the three R's. Now, perhaps they would have better minds if they learned the Greek letters and language instead, but they have no choice. This English is their heritage. Long before they could choose, their elders have moulded their minds and made them into English speakers, English readers, English writers, and accountants. The young depend on the choices made for them by their elders. An heir is not somebody who can choose what he shall inherit; if he could make his choice, he would be self-made. But, in so far as his inheritance is determined, he is an heir, and under the laws of heredity. And the first and last word of heredity means that his background is given. He does, however, determine the background of the next generation.

Hence, one's generation's background is due to the previous generation's foreground. My father's values determined my education. And by no action of mine can I cancel out the fact that his education preceded my own judgments. I am more the product of his intent or his omissions than his own life was. I am his heir. Only my own son or students may fully reflect my own choices.

Society is based upon a principle of dovetailing which is unknown in the animal kingdom. To ourselves and to the education of ourselves, we arrive too late. The most important effects have already affected us when we come to think of them. Of course when they have done their work, we may reconsider and doubt them, and act differently. But, since we ourselves are already determined, our new conclusions stand a fairer chance of bearing fruit in others than in ourselves.

IV.

This constitutes the great human secret. Mendel's mutation takes place in the conception and birth of the individual. Our historical mutability, on the other hand, is effective as a mental relation between two people, two generations, two times. Those qualities or energies which link at least two, and weld them into a cooperative being, "transpeciate" our species constantly into new men; naturally, these qualities can only be found when they are not looked for within the individual. In him we shall never discover how any social function unifies the speaker and the doer, the first and the second doer, and so on. These energies must be processes between two minds, two hearts, two people, at least; perhaps between many more. The obstinacy with which psychology has studied the mental processes within the individual, is no proof that its method is fruitful. The dream of a self-taught, self-ruling man is a bad dream. The measure for teaching and ruling cannot be found from the abnormal compression of these processes into one individual. Historical man is taught by others, and rules others; and in these relations, he is compelled to realize himself. "He," never exists, but is always between two times, two ages, as son and father, layman and expert, the end of one era, and the beginning of another. In their despair, the mental monadists—who look for the mind inside the individual—call our time a period of transition. Sheer nonsense; the essence of time is transition. In so far as we act or speak, we can act or speak meaningfully only between two other generations preceding and succeeding us, because we always come too late to ourselves.

My self is not the container for my acts and ideas. My acts carry out the ideas implanted in myself. My ideas plant the acts in somebody else because he will be purer to receive them. Since this is so, our will is not a vehicle for making ourselves. Freedom of the will is not the subject matter for self-worship or self-reliance. Freedom is given us because of our functions as enders and beginners. Our function as children requires to be superseded by our function as parents. The child is certainly not father of the man; I think that this is the central fallacy of mental theory. The romanticism of Rousseau and Wordsworth destroyed the continuity between generations; and as a substitute condemned the poor children to carry a burden which rightly their elders should bear. The child prodigy of the 19th century is the ghastly result of this impatience with the individual. He was in a hurry to be his own father; as a reaction, he usually remained childish. *Freedom is given us for the race.* If we try to interpret freedom as given us for ourselves, we grossly exaggerate our abilities. If we deny freedom, we fall into the snares of racial servitude. Jesus remained under the Jewish law to the end of his first thirty years. It took him the time span of what is called one generation, to outgrow the synagogue. His obedience consisted in his patient walk through life. The risen Christ may walk with all men. He could not belong to the ages if he had gone at nineteen into the desert and founded a sect then and there. In his walk through his generation if we walk with him, we are all freed from our native limitations.

It is the whole content of Christianity that we are free, but that we arrive too late at our own freedom for fully wielding its liberating forces ourselves. Our own time is a station between the times which our freedom rejects and the times which our patience prepares. The meaning of liberty is our power of creating a new kind of man. This power is capable of curing the breach between the mere fashions of each separate generation. Enslaved by the latest trend and current events, we rush to the worship of the gods of our days. And these gods of the day follow each other in a cycle similar to the business cycle. Every season of the mind automatically begets its undertaker in [the] form of the opposite philosophy. And in the cycle of all possible philosophies the poor devils are caught blindly.

But we can wake up and see the cycle and break its spell and create peace beyond the warring spirits of the times. This power was the distinguishing feature of our era [i.e., the Christian era]. Therefore when this power goes into eclipse, we are back to paganism, immediately. And in paganism, eternal war is the order of the day, and war only, between all the spirits of men. Accordingly, the Nazis who proclaimed eternal war, and "annihilation" and "elimination", banked on the one generation of the "Youth Movement" which broke away from all peaceful relations with their parents. All the revolutionaries in Europe are "matters of one age" and play up their own spirit ruthlessly. The Nazis reveal that if one generation may carry out its temporal spirit unhampered, war becomes the only principle of life.

But it is no good to retreat, in the face of this relapse into paganism, into the eternal city of peace. The eternalists would like to look down with superiority into the arena of human fighting. We have heard their protest of an eternal peace, and certainly the Pacifists are the indispensable antithesis to the ghastly warhoops of the temporal mind. But the antithesis is Pharisean and incomplete.

They are right when they abhor war as the order of the world; it certainly is its disorder. The world was created for peace. But they are wrong when they do not add that the act of creating the world is a perpetual act. What we call the creation of the world is not an event of yesterday, but the event of all times, and goes on right under our noses. Every generation has the divine liberty of recreating the world.

The pagan mind stares at war, finds it everywhere and proclaims it the form of life. The eternalist stares at peace and proclaims it the content of life. Both suffer from a fixation. Both are lacking in freedom. The soul knows that we move in a world at war to bring peace into it. In every hour of history the recreation of that peace which was created into the world as its goal from the beginning, is the topic of our fight. Between the war party who places itself on the side of the world "as it is", only, and the peace party which places itself on the side of God only, our loves, hopes and beliefs force us to proclaim a "war and peace" party. The Cities of Men and the City of God form one crucial unity in a living person. The chaotic world at war, and the emerging new peace for this chaos which made the war, are the two aspects of a mankind in cooperation.

V.

At this point, the majesty of the educational vogue of the last hundred years becomes clear. Liberalism as mere anarchy of beliefs or values does not impress me. But liberalism as willingness of parents to give their children a futuristic education, strikes me as great. These parents were ready to let their children go further than they themselves could reach. The true Christian spirit of Liberalism lies in this willingness of whole generations to let the next generation go into a future precluded to the parents themselves.

Since the liberal anarchy of standards for the individual around us is so colossal our fundamentalists easily overlook this very definite creed of the agnostic age. Between the generations, a bond of parent's love and children's faith was established which translated the parents' hopes into the children's lives. This should make us feel reverent.

However, although the parents made the sacrifice, the institutions of learning did not do the same. When the boys came to school and college, the older generation declined to mould them into new men. The transpersonal energies which connect two ages, were denied. Man had, they were told, his own mind to and for himself. And so the teachers and students on our Campuses lived under the fiction that they were contemporaries and could feel and think the same things. Nobody was responsible for anybody else's thoughts; nobody was meant to be his brother's keeper.

At this moment, a young generation is in a new kind of war, a war which is not based on a settled society of the past, but on an industrial society of constant flux and change. This generation does not fight as all former patriots did, for the father's laws and order, because they know from their fathers themselves that change is of the essence. Change, so they have been taught, is their birthright. So, if they shall fight, the soldiers must fight for a future beyond the war, not for the past as it was before this crisis.

Our soldiers wait for opportunities, not simply of returning home, but of turning towards a new peace and of immigrating into the future. The morale of this army will depend largely on a change of heart in the articulating generations, the people who teach, write, speak, and occasionally think.

Our schools have tried to teach the boys and girls the values which we feel they should think. That usually meant that they were asked to feel that which we thought. So they ceased to feel. Now, the discovery of the two-generation-way of the mind in action, involves a tremendous change. The young first must be allowed to feel, to scent, to presage, to fight for themselves, to quench evil, to protect the world before we can speak to them theoretically. The old must think out lucidly that which the young, have felt or can feel about the future. "We may conceive humanity as engaged in an internecine conflict between youth and age. Youth is not defined by years but by the creative impulse to make something. The aged are those, who before all things, desire not to make a mistake. Logic is the olive branch from the old to the young."[2]

In other words, the thinker (any man, old or young, who is asked a question, finds himself in this awkward role) should not ask the doer (the man who is about to act, perhaps on the basis of his answer) to share the detachment of the thinker. This, however, is what our academic education does, and the detachment of the thinker-answerer is recommended as the only right emotional climate. "Don't get excited"

---

2. Alfred N. Whitehead, *The Aims of Education* (New York, 1929), 179.

is no wise counsel to young men. If they no longer can get excited, the world decays, just as much as when the old men can't keep cool.

Therefore, the thinkers should try to think out clearly the same processes which work up the emotions of the soldier of life so deeply that he is willing to give his life for safeguarding order. The thinker's clarity should match the soldier's intensity, without ever forgetting that by his clarity he tries to rival the heat engendered in a human heart. Therefore, the thinker depends upon the flames of passion burning in the doer, and these high temperatures provoke and challenge his effort in lucidity and dispassionateness. These flames must burn without smoke.

The collaboration of soldiers and thinkers must be the central article of any society's constitution. Only then, will the thinkers drop all pettiness and rediscover those truths which are vital.

VI.

This would be nothing but applied Christianity. It would carry the evangelical relation which has grown up between parents and children during the century of Liberalism to its logical conclusion. And the schools and colleges would now undergo the same conversion which the physical parents underwent when the Autocrat of the Breakfast Table was buried.[3]

When the parents ceased to play God Almighty for the beliefs of their children, they did something of significance for the universal relation between thinkers and soldiers: they trusted the young.

Must not anybody who is asked a question about the road, and has to find an answer, speak cautiously, trying to make no mistake because the other fellow might march off wrongly upon the answer? The soul's delight is in doing this, here; the mind's genius is to think all in the proper system. As cautious as my answer to the stranger who asks for directions, as bold must be the action of the man who goes to war. Thought is born by circumspection. But a soul is born through the growing pains of suffering in action.

Before this war, our schools have preached to the young to avoid conflict, to avoid pains, even the growing pains of that suffering which

3. *Ed.* Oliver Wendell Holmes, Sr., *The Autocrat of the Breakfast Table* (1858), a collection of essays

is the inexorable counterpart of acting according to your heart's command. When Aeschylus said that the counsel of Zeus prevailed, which ordained that the man who acted had to suffer, he said something as true as that "two and two is four." But this interplay of action and "passion," doing and paying the penalty for it, has been ridiculed by sociologists, psychologists and all the monadists.[4] And so, they sterilized the young.

Arthur, the asker of the bold question, "Is not the Old Testament a bundle of paganisms," may set us right. He came out right into the open trusting his instinct, and gave battle; he took the risk of blundering. However, he also took the trouble of exposing this impulse to criticism. He trusted himself and he trusted his older friend. In this double trust, there could grow up a new answer which his generation could understand. If the two trusts could be made to grow, both till they embraced the whole of life and truth, the people would build the bridge of real time again, after a lag of thirty years. For thirty years, the Holy Ghost had abdicated in favor of the Spirit of the Times and the wit of the individual. The thinking of our college men became childish because the old and young tried to obliterate their difference in age, and played together as though they were of one age. Compromise became the great slogan. Before they probed into the depth of their feelings or the profoundness of their thoughts, people hastened to compromise. And these compromises satisfied as much as did the Missouri Compromise; it did not create one common reality between different generations.[5] Since nobody took the trouble to pour his real desire or his real vision into the compromise, no promise was fulfilled by it, and all hope was frustrated.

When my friend Arthur understood this, he wrote me one sentence which touched me to the quick. I had not foreseen this reaction; and I still stand in admiration before this lucid sentence: "Oh," he wrote, and this is a literal quotation, "I am a pagan; for, I have no speech."

He had discovered paganism to be the lack of relation between the generations of mankind. And in the process, he had made the much more important discovery that speech is not the byproduct

4. *Ed.* "Passion" here means "undergoing," as in "passive".

5. *Ed.* Missouri Compromise of 1820 was an effort by the U. S. Congress to balance the interests of both slave and free states.

of individual action or individual thinking, but that we speak with power only when acts and thoughts meet. Our tongue, our power of the word to which the millions obediently march and serve and sacrifice, is not the "expression" of scientific ideas, or the war cry of blindly marching cohorts. The living speech of a community results from the polarization of acts and thoughts; like the spark which crosses the dark gap between the positive and negative pole of electricity, speech is a flaming arc, connecting different generations. On the one hand, blind acts are speechless, and who does not know the dumbness of the mere busybody? But—and that is mostly forgotten—similarly, abstract ideas are speechless; in a sense, all science is nearly speechless; it is a whisper between experts. Only when *taught*, only when facing a new generation, does science recover speech. The blessing which results when thinkers and soldiers face each other is that public speaking is reborn. Speech blends the two processes of pure thought and pure action. Arthur found himself outside this electric arc; but whereas most of us remain unconscious of our exile, Arthur discovered that we have, as children and parents, a spiritual office, in the never ending chain of generations. *The links of a chain must overlap. The evolutionary scheme of the last century omitted this big question of overlapping*, of putting the rings of the chain together. Lest the chain remain unforged, children and parents cannot behave as though they were contemporaries. Both must go to the edge of life, in militancy against danger, for the reorientation of the species; one exposing their physical life as soldiers, the other exposing their social reputation as thinkers. (This is the reason why no progress in human thought is possible without the martyrs of thought or science.) A brave man is he who risks his status quo lest new life be stifled or higher life be destroyed. When we eat, when we breathe, we integrate lower life into richer life; our social acts obey the same law. The physical existence, in the soldier's case, and in the thinker's case, the moral existence, are the chips which we stake for the essence.

By now, the soldier will be seen as a spiritual agent, while we come to admit that thinking itself is risky action. The "spirit" is a comprehensive term for both, action and thinking. When the spirit of France died, she lost Paris, her intellectual center, and Toulon, the center of her imperial strength, both. To the spirit, mind and body are both mere matériel.

If there is one spirit, thinkers and soldiers move in one common sense. And there is one spirit when the parent- thinker in us brings up his pearl of thought out of the same darkness in which the son is plunged by the feelings and passions of his youth. We should think up to our impulses and feelings, not as it is the fashion, drag our lives behind some abstraction, some "ism," of our mind's making. If a thinker rethinks the truth in the light of a doer's vital impulses and actions, the future way of life lies open again as it was proclaimed in the beginning of our era as the good news. New lives may be lived in freedom; the young may trust their vital instincts, no guilt from the past shall asperse them, for their elders will forge an armour of thought around their heart's flames. The expert may retranslate this into theology. For my naval "JG", the quotation of "original sin" at this point would add little.[6] He is impatient to see the right relation restored: "I feel certain you won't let us down."

Thinking for soldiers, instead of ruminating for children, is a very new aspect of research and education. But this is the reform of our educational system which the three witnesses demand; the speechless college senior, the "forgetful" lieutenant, and the man who leaves the Church when he enters the army. Higher education in the future can only be planned for people who serve and fight life's battles, on whatever fighting front.

Otherwise, the bodies of the young might be slaughtered for the dated ideas of a senile science, or the mature ideas of truth might be butchered by the rash instincts of brutes. In between lies the road of atonement between the body of young life and the mind of old life. These two have to coexist and to interpenetrate.

In this, what else do we say than that which was known always? The coexistence of more than one generation at the same time, the deliverance from blind cycles and sequences, was called the achievement of the Holy Spirit. He was conceived as proceeding from the Father and the Son. We all know that a father's mind should enter into the impulses of his son. That is the reason why nobody may call himself a father, by mere physical procreation. Fatherhood is rethinking the world in the light of one's children. Why is God so inexhaustibly original? Because he rethinks the world for every generation of his children.

6. *Ed.* "JG" is short for "junior grade" as in "lieutenant jg".

Beyond the level of brains or brutes, of scientists and warriors, the soul is born. The soul in the scientist makes him into a teacher; and the soul of the warrior turns him into a soldier. All through this essay, we spoke not of "scientist" and "warrior", but of "soldier" and "thinker", because all the time we anticipated a mutual recognition, between the representatives of war and the representatives of peace. They are brothers. They can speak to each other. And this is the inspiration which was promised us as the Comforter through the ages.

Five

THE METABOLISM OF
THE SCIENCES

Ca. 1945, Published in German in *Der Atem des Geistes* (1951),
167–235, and in *Die Sprache des Menschengeschlechts*, II (1964),
221–275.

INTRODUCTION

Heuristic means creative or inventive, in the life of the intellect. Eureka, I have discovered, contains the same Greek stem. Euristics or heuristics formed a recognized field of philosophy. Today, the sciences have no common consciousness left. And the very term heuristics for their common method of producing new sciences and new methods within one single science, has been forgotten.

The progress of science depends on the reconquest of this Heuristic principle. Otherwise, the terms which every science uses and from which it takes its start, remain accidental. The sciences lose their power to prune their tree or branches as soon as they go positivistic. Because scientific positivism means that scientists decline to discuss the conditions under which any new science springs from the life of society, scientific positivism makes science into something for which the scientists need not render accounts to anybody. Scientific positivism expects that everybody in the community will contribute taxes and donations for the sake of science but that this naive faith of the people in science otherwise may be taken for granted.

Scientific positivism and the disappearance of Heuristics, then, are one and the same event. From 1870 to 1940 we may say that science cut the umbilical cord which connected it with the religion of the community, and declined to discuss the proper relations between folklore, science, and slogans.

But science is a halfway-house between the lore of yesterday and the slogans of tomorrow. "Evolution" was a victory over folklore when it came into scientific use. It is a derelict, a slogan of the educated mob compared to today's scientific problems.

René Descartes, the father of modern science, called it a half-way-house. We propose to take this expression up once more and to expose its location[,] between two ends of a road[,] to the sight of society. If somebody lives in a house halfway between two towns, he may dream of being alone in the country. Not until a bridge becomes unpassable or foreign soldiers maraude, does he wake up to his HALFWAY-HOUSE reality. Something like this is happening to many good people in the palace of science because of the atomic bomb. They do realize that their home stands in some community. But they are quite helpless to determine the relations of their spiritual home, science, and the community. For what is a community. A community is the group in which lore and mores, sciences and techniques, slogans and politics, move in a perpetual interplay by the compelling force of common speech and language.

The compelling character of the term NATURE, for instance, bound the Nature of Man, the Nature of God, and the Nature of Science into one functioning triad of lore, research, and slogan. Nature of Man today must be dismissed as inept a term. Society is man's secret. He has no NATURE. Immediately, the natural sciences are dislodged; the one term the community shared with the scientific brotherhood has collapsed.

The compelling character of the speech that the community instills into the future scientists, and the compelling character of the conventions that the scientists formulate, on the community, are the topic of this paper. The interaction of science and the people is not a material interaction as though the people "benefit" by science. The story is much more the story of a delicate dynamic balance between the forces of community life on the levels of an integrated community, an uprooted intellectual group, and a rerooted society.

This interaction is based on the vicissitudes of speech in the community, in science, and in politics. This is the lifeblood which runs in the arteries of the communities, in the veins of politics, and through the lungs of science. Each time it is in a different state.

The scientific positivists have never allowed themselves to pause and to ponder over this metabolism of speech through the medium of the sciences. "What of it," they would say of languages; "it is a poor instrument anyway. Our thoughts suffer from the illogical character of human language. If nothing but speech ties together 1) the community prior to science, 2) the scientists, in their halfway house, and, 3) the enlightened society, after science has done its work, then it is not worthwhile to look into the matter. What is in words."

The misery of our present state appears in this horrid phrase: "Nothing but words; nothing but speech."

Speech is everything. The health of science, the health of the community, the attainments of society, depend on the recirculation of speech between the two ends of the road and the HALFWAY HOUSE of the sciences. This, then, is the theme of our paper, and, therefore, it is called THE METABOLISM OF THE SCIENCES.

THE NATURE OF THE PHYSICAL WORLD

"The Nature of the Physical World" is the title of a well-known volume of Gifford Lectures by the physicist Eddington.[1] It is an elegant title. For in it, the social, religious, political, and mental issues straddled by science, are reflected as by a prism. Innocent and scientific it may sound; in fact, it is past innocence and this side of science. To determine its place in our society, is the aim of this essay.

We shall see that the religion of the physicist stands revealed and not only of the physicist. The religion behind all science stands revealed and the religion which these scientists share with the nations of the world. For on the basis of this religion the nations allow and demand that a universal science shall operate right across all their

---

1. *Ed.* Arthur Eddington (1882–1944).

political borders. From 1440 to 1946, that is until physical research came under government control, science was international.

That a book title is so pregnant with meaning, is rare. But that a book title in itself is symptomatic of the faith in the community in which it is published, is to be expected. Books hold the position of children of their authors. In naming our children, we cannot help declaring our faith or unfaith. Eugen, Amos, Baldur, Harold, speak on the monumental level of life-long names. Our words may be of the moment. If I call my child Trifle, I certainly betray some definite cynicism about the value of a human soul. Because, our names ride on the wavelength on which more than one generation overlap. The name which I give must be valid in the spirit of my own time and in the spirit of the child's lifetime, and finally, in this child's children's lifetime. Now whenever we declare ourselves in the face of more than our own generation, we are compelled to disclose our religion. In our own time, we may put our light under the bushel and may conform. Between the spirits of many generations, we must become emphatic and are found out with regard to the things in which we really glory. And a man glories in his faith or in his cynicism, in his personal or in his conventional faith. But glory he does when he must represent his whole age and its spirit in the teeth of other unknown generations.

Hence, names are the declarations of our faith whether we like it or not. This being so, *The Nature of the Physical World* declares the faith in which science and the nations of the Renaissance world glory. The average academic reader already at this point may rebel. He knows how book titles are fabricated by publishers. The irreverence of the commercial book market now fills the academic spectator with irreverence for all book titles.

May I suggest that this, though clever, does not seem clever enough? The objective observer of the wiles and tides of book titles may combine complete contempt for mere salesmanship with the utmost reverence for the stream of speech which the publishing craze pollutes. It is the academic mind's curse that it will not revere the very names which it must use in order to be itself respected by society. And yet, "science," "Plato," "truth," "happiness," "greatest number"— all these words are indispensable for the very existence of any science inside our ruthless society. But let it be understood at the outset that the flippancy of the enlightened reader with regard to book titles may easily [~~devise lines~~; *illegible hand written correction*] of an access to the workings of his own mind. For he too, opens books and looks at books and writes or plans books or is in search of books treating certain questions and not all those are free who boast of being without roots in or ties to the living tissue of language.

Hence, I cannot help it if right here some readers will part company with me. There is today a large group of scientific and literary men who are eager to scrutinize any footnote and any word or term used inside a book, and who, nevertheless, will treat my thesis as absurd that book titles are the clearest expressions of a society's religion. These men point to the racket of catch words, of slogans, to the arbitrary inventions of book titles by enterprising publishers or agents. And they actually think that the abuse of our good faith in book titles refutes the right use. Actually all the facts of which they complain prove my point. *Corruptio optimi pessima* [corruption of the best is the worst]. The most important elements of life are of course most often falsified. We will discuss these abuses at a more advanced stage of our investigation with much greater ease and profit.

But it seemed wise to bid farewell right here to the sophisticated people who no longer are able to see what they are doing when they themselves read the New York Times Book Review, mostly without ever investigating the verdicts of this oracle or reading the books whose names are listed in it.

With this group of readers unfortunately lost, I return to our serious issue that "The Nature of the Physical World" is the very statement in which science glories.

And since I now have ascertained our right to take this book title seriously, a very startling admission may be added.

Books with this and many similar titles have become part of our religion. They are expected by the public, they are desired and they are written. They are part of our living faith. To believe in our right and our duty as well to produce books on the Nature of the Physical World, in other words, is a part of our religion since the Renaissance.

It is a vital part of the living faith in which the Cardinals who fly to Rome by plane and the Japanese suicide flyer find themselves united. On the other hand, it is only a small part of the religion of a Cardinal or of a Japanese. And this is the exciting truth that the belief in a Nature of the Physical World can never be more than a part of our faith and on the other hand can never be treated as anything less than a religious act.

We are today attacked from both sides, one side claiming that one may have science although society treats it as a commodity, the other side pretending that the faith in science is all that is needed. Science today is in danger of decay because some idolize it and some look upon it without religion. Sectarian science and commercial science are equally ugly, and equally destructive of scientific progress. The most orthodox Churchman today must take pity on these dangers to science. And it is as an orthodox thinker of the Christian dogma, that I wish to defend the religion of science today as a vital part of my religion, against its abuse.

I wish to develop the faith of Renaissance Christianity out of this book title, The Nature of the Physical World. For this purpose, I shall take the following steps.

First, we shall analyze briefly the name of the book. Of this name, we may get hold in one glance. Second, we shall look upon the work sheet of some hours of laboratory work. On it, the things a beginner is doing from respect to the Nature of the Physical World, can be seen; things which the good man does to himself[,] by the way[,] since all faith makes us do terrible things to ourselves. The work sheet shall reveal to us the neophyte's self-immolation. Third step: The state of mind of a mastermind over decades shall become pellucid through extracts from Faraday's daily notes.

These three steps unfold, for widening units of time, the living faith in the Nature of the Physical World. The fourth step will be to discover parallels to this name of our book, in all other fields of science, and to deduce the common law of their formulations. We shall compare this linguistic pattern of the last four hundred years of the Renaissance with the religious pattern of antiquity. And at the end, we may understand our own faith somewhat better, a faith which produces such an exciting, mysterious and absurd title of a book.

For let it be understood at the outset that it is exciting, absurd and mysterious, all three. The Nature of the Physical World? Aye, why not the physics of the natural World? or The World of Physical Nature? We may well ask so naively. For, the three terms composing the title, actually say three times one and the same thing. One "X", so to speak, is labelled, whether we call it "World", "Physis", or "Nature". Physis in Greek is the same as Nature in Latin and World in Anglosaxon. If we should give a definition of this "X" behind all three terms, we might say, the world, nature, physis, are different expressions for the universe when we speak of it and treat it as speechless.

But then, the treble repetition is an obvious mystery. Perhaps it is not for our blunted academic sensibilities which are easily imposed upon by Hegelian pomposities. But for a singer of the wild, it would be only too obvious. He would immediately compare our title to the magic formulas known to him, for the obvious reason of its being repetitive. Any magic formula operates by emphatic repetition in such a manner that the very fact of repetition is in itself a part of the formula. Usually, the Open Sesame is said thrice. In Macbeth, the three witches sing: "Thrice to thine and thrice to mine and thrice again to make up nine". A Roman prayer, in fact the oldest Roman prayer preserved and a descendant of the Greek pattern of prayer at the same time, is built in the same manner of threefold repetition. As we shall have to say more of this prayer later on, its full text is given in the appendix. The power of the thrice repeated lines, [Hermann] Usener [1834–1905] has called this the world-wide tyranny of trebling, is apparent right through it.

"The Nature of the Physical World" speaks of the same thing, of the speechless universe, thrice. The Nouns used are equivalents. Herakleitos the Greek said of physis the very things we say of the World and Cicero or Lucretius said of Nature. In as far as the book title is repetitive, the formula is magical in its appeal to the public. But you and I know that it is not black magic which is offered us. It is legitimate white magic, alias science.

Then, the saving grace must spring from the alternation between the three linguistic layers, Latin, Greek, Anglosaxon. And this indeed is the case. We do not accuse Eddington of an unduly superstitious appeal nor do we accuse science of being black magic. However, we insist that whenever science invites a legitimate and yet not scientific public, the relationship of science to magic is undeniably conspicuous.

However white science appears to us[,] it retains a definite though antithetical relation to magic. It is an incantation and it casts a spell. You may say that it is a legitimate incantation and these spells are desirable. But this is not under debate. I would say that science is verified magic, magic come true. Just the same, it is important that, we have admitted this verified magic into our ken.

By saying this, I have already stated that science has been admitted by the children of God despite the fact that God has not created a "physical" world and that we as his children, know absolutely nothing of such a physical world. God created the world; he did not create the physical world as a world by itself. If this should surprise you, you merely have to analyze now the second quality of our book title: its absurdity. If one world is postulated as "physical", another world, which is non-physical, is admitted, too. The one adjective "physical", limits the world which is thus qualified. And immediately, the mental world, the political world, aye, finally even this strangest term of a Christian World, rise before us. The Renaissance mind is seriously impaired by this divorce. Scientists often look down on people who speak of God's face or God's finger as being hopelessly superstitious while they speak of God's mind, themselves. But the mind of God is as much a metaphor as his elbow. Our mind is not nearer to God than our body. Yet this division of the world into a physical and a mental has blinded many as though the mind were more divine than our kidneys. To the fundamentalist thinker within me, my acquiescence in a physical, mental, Christian world is absurd because by virtue of my orthodoxy, I hope to believe in the fact that God created One World which includes all of me, mind as well as body. But to my "re-search" mind, the split is a natural!

Therefore, I find myself compelled to retain both positions. It is true that God created One World inclusive of minds as well as bodies. On the other hand, the absurdity that there is a special physical world, is at the bottom of all science; and we, that is to say the peoples of the Christian nations, have become persuaded that we should allow ourselves to split the wholeness of One World into a physical and a non-physical world. We have become convinced that ultimately we shall benefit by supporting the absurdity of separating a physical from a mental world, and of making the one, the physical, into the object of the other, the mental.

This then is the exciting paradox revealed at first glance to us by the mysterious and absurd title, that we have admitted into our society a process which contradicts the first article of our faith, a process which shares its formula with all magic.

Now we take the second step. It leads us into a laboratory. The work sheet which I reproduce—I myself assisted in this work during the war—reflects the physicist's self-immolation on the altar of science. With the worksheet, the reader so to speak, holds in his own hands the dichotomy of a physical world which has become the object of a non-physical world, and of a mental world which has become the subject of the objectified physical world.

OUR FISSION IN MIND AND BODY

We shall observe the process by which at the end, two fruits are produced by faithful research which do not exist in reality but merely in experimentation: a new subject and a new object as the world has not seen.

The reader finds that the worksheet is divided into two pages. On the page to the right, he finds the term "data", with (A), (B), (C), (D), (E) neatly listed. On the left page he finds scrawls. We will now please concentrate on the difference in style between the two pages written concurrently by the same man during one and the same experiment.

We are interested not in the experiment as such or in its result but in the interplay between the two pages. The right page is employed for "data", the left for figuring. In this, two processes are set in motion, the right page processes towards creating objective result, the left page towards creating one unified subject mind. The data may be called data of *observation* since the term data means observations made by the five senses either on an instrument as to its readings or on matter as to its visible, audible, smellable, tasteable, touchable behavior. The figures are immersed in a process of *computation*. They are added and multiplied, etc. by "arithmetic" or "algebra" depending on the necessity of computing either figures or letters. As the handwriting shows, the observer meets his sense data with a firm hand. He faces the facts of the experiment as one individual who meets other individuals. His statement is definite, his form of writing well defined. He stands at attention like a soldier on guard, fully equipped with his faculties of keen observation. But there also is caution. The reader finds under (A) and (B) that 3 or 4 different readings are listed. As a soldier on guard does not shoot before he has asked several times, so the sense data must not be guess work. Science in an experiment repeats the readings. By this precaution, modern physics reverses the process of magic. In antiquity, the word or formula would be repeated three or four or seven times to make sure that it did catch the natural process. We do not repeat the magic formula, but the observation. We do not suspect the reality outside but our own senses. We check and recheck our data. The three or four readings of one and the same phenomena check our sense report of the external world. Hence, we have the right to say: One observation no observation. The isolated data is still "pre-objective". Only a series of data leads beyond mere impressions. Not one impression is a real datum; the whole list is one. Hence, the reader of our sheet finds behind all these sequences of data a ±.005 or ±.01 cm (.39%) is the stigma attached to the senses. Pre-objective impressions never are perfect. A margin of error remains. And by this margin of error, the whole list of observations is still off the ideal mark of perfection. Ideal science can only deal with reversible process which can be repeated. For, in an experiment, unique processes can never be objectified.

Three steps are taken: a single impression on a sense of our body, a sequence of such impressions is formed and an average is taken with a possible error of so and so many per cent. This means that the primitive single sense reaction is thrice removed from the real standardized observer in us.

Now, we pass over to the opposite page. At first sight, the style stands revealed as dramatically opposite. The hand which was so definite and firm on the right page, has moved hastily and nervously. It is not tied down by the horizontal and vertical lines of the sheet. It runs in more than one direction. Most computation sheets look even more ghastly and criss-cross. Is this accident? We do not accept this escape. In style, anything spontaneous has the weight of substantial truth, of a telling fact. Any work sheet, by the way, in hundreds or thousands of cases, produces the same effect.

Therefore, we have the precious evidence of the grammatical dualism which is operative during the period of suspended judgment. The left page is the subjective page. Computation is a purely mental activity. And because it is, the bodily phenomena which accompany it, show not a man on guard and at attention, but relaxed, indifferent to appearances, introvert in slippers. For the grammarian, mathematics is not a question of content but of form. What is computation doing, in this wholly introspective process on the loft page?

We find, for instance—the reader need not fear that he has to figure it out himself—the equations

$$R = \frac{.2525}{2} + \frac{(2.57)^2}{2(2525)}$$
and
$$R = 13.20 \text{ cm} + .13.$$

What does this represent? The two fractions of the first equation, one at single power, the other to the square, have been reduced to one expression. Well, "what of it", the objection may come; "this reducing is our daily bread". But this daily bread of physics by which he reduces different expressions to unity, is as admirable and as mysterious as real bread. Could it not be that the daily routine blinds us to the transformation's full significance?

Something went overboard, for unity's sake, in the reduction. It was treated as ballast. In this special case, it was, among others, the term "to the square" in $(2.57)^2$. To reduce to a common denominator means to sacrifice a nomen, an expression, a particular name. Mathematics redefines its expressions so long and so consistently as to achieve the greatest possible unity of expression. The mind on the left page, sacrifices expressions, and there may be as many as a hundred of such mental sacrifices, on one page.

What is the net gain? The subject who at the end, says 13.20 cm, although in his private life he speaks of inches, has sacrificed his historical vocabulary and nomenclature. By this purification, he has become one mind with all other people who compute, all over the globe. His mind now is a so-called transcendental ego, a mind detached from place and time. Place and time always are limited by names with their local and temporal associations. The transcendental Ego who emerges from our left page, is detached from his native and social attachments, and owes allegiance only to the republic of physicists. In this republic, a special language is whispered, the language of mathematics. This language because it is a secondary language, is not spoken but moves by signs. Also, being secondary, it is nameless. But it is meaningless unless it emerges from a primary layer of speech. Mathematics on the left page must receive something which they can reduce. Or, there is no room for its proceedings.

Let us assume that on the right page the data were partly measured in inches, partly in centimeters. On the left page, we then would read the reduction of inches to centimeters or vice versa. In this manipulation, it would be obvious that one expression, "inch", or "centimeter" was sacrificed to the victor. But the expressions ·2525/2 and $(2.57)^2$ are two "expressions" in no less degree than inches and centimeters. All expressions are Names waiting to be sacrificed in the quest for unity. If they were left to themselves, they would remain *inaccessible* to each other. We have to reduce them by *cutting off the head of one* of them before they can be incorporated in one statement: Computation requires amputation. Why? By these amputations, the scientist becomes one mind with all other people who compute, all over the globe. The mind that emerges from those amputations owes allegiance only to the international republic of science. Incessantly, computation sacrifices expressions for the sake of unity. I have given the history of the decimal system which was introduced by the men of 1789, in my book *Out of Revolution*. It certainly is a most dramatic conflict between two sets of values, one scientific, the other social. However you side in it, it is a real conflict because names deserve to be kept, at times. Reducing Dante and Milton to a statistics of their verbs, in the data of their poems, may be valuable. Mario Praz has reduced d'Annunzio's famous poem *L'onda* to a string of quotations from the Italian Dictionary which the "poet" simply had versified.[2] But these reductions are irrelevant in genuine poetry for the simple reason that in poetry, the names are relevant.

For the sake of unity, I split. This sounds like a contradiction in terms. Yet it is true. The allegiance of my mind to the republic of computation in which we are all of one mind, and the allegiance of my five senses to the sense data in which they are immersed, produce a rift in me. Because the experiment takes time, the time in which body data and mental reduction, resist each other, our attention is drawn to this conflict of body and mind. And when many men undergo such experimental training and experience, they begin to divide their own

2. *Ed.* Mario Praz (1896–1982), Italian scholar who wrote influentially about English literature.

being *into body and mind*. But they mostly overlook the fact that it is not their own body and their own mind which are separated, in the scientific process. Our two pages tell the true story of this division of mind and body. It is perceptible only when the mind enters into the fellowship of other minds, and when the body bends over and attaches itself to matter, to other bodies, as their pure organ of sense registration. Man, in a scientific experiment establishes two specific fellowships: one for his senses, one for his mental powers. Truly suspended like Prometheus on the rock, the scientist may not descend from his suspended and outstretched position between sense data and computation before he has not bent the two worlds of mathematics and of material physics to each other, through his power of reducing one to the other. Mind and body are means to an end. Man divides himself into mind and body, temporarily, for a specific purpose. And the purpose is to force the world of the senses to admit of a unity in figures. During the suspense of the experiment, the mind all the time becomes more of a mind; the body the more faithfully we observe, becomes all the time more clearly the body. Objects and subjects do not exist, but are polarities produced in the action by which we split inside temporarily for the purpose of uniting afterwards more consistently. A given diversity and a desired unity contradict each other in the beginning. The physicist undergoes voluntarily, for the sake of solving this dilemma, a cleavage inside himself. With his body he forms an element of the physical world, with his mind, he forms an element of the mental world.

The division in Mind and Body, in a mind of mine which is a part of The Mind, and in a body of mine which is immersed in a physical continuum of bodiliness, is not a natural fact of our existence. To the contrary, it is a sacrifice of our personal destiny for Oneness, Wholeness, Singleness. A bridegroom, a soldier, a daughter, must ignore this division, lest mankind perish. Their whole being, this side of any such separation in mind and body, must stay in organship to creation.

Science requires the sacrifice of this naive organship of the creature man so that he may instead become instrumental for the construction of a polarity; in this polarity, his unity is pulverized between the two foci of an ellipse as all his observations push him one way and all his computations the other.

THE SECRET OF THE SCIENTIST

We have explored the style of the worksheet for one experiment. Let us now ascend to the next level of time, to the life work of a physicist over a number of decades.

In Faraday's many volumes of daily entries we have access to the whole life work of a great master. The last paragraph of the seven volumes of Faraday's daily reports on his work bears the number 16,041. And one of his last public utterances was: "For all the phenomena of nature lead us to believe that the great and governing law is one".[3]

16,041 and One, Unity against the ocean of 16,041 data, this is the polarity of his grammar. Both poles are stressed and should be stressed. Tyndall said of him: "A good experiment would almost make him dance with delight".[4] Yet Lord Rutherford also could write: "When we consider the life work of Faraday, it is clear that his researches were guided and inspired by the strong belief that the various forces of nature were inter-related and dependent on one another. It is not too much to say that this philosophic conviction gave the impulse and driving power in most of his researches and is the key to the extraordinary success." (Report on the Faraday Celebrations 1931, 39).

Because Faraday spent his life suspended between the One Nature beckoning from the future, and his daily 16,000 experiments, it was written of him: "The Contemplation of Nature and his own relation to her, produced in Faraday a Kind of exaltation".[5]

The diaries establish this respiratory process of the mind in action between 16,041 reasoned out doubts and the foreshadowing of One Faith. And the quotations which now follow are not more than a few examples.

"Surely, this force of gravity must be capable of an experimental relation to electricity, magnetism and other forces, so as to bind it up with them in reciprocal action and equivalent."

"I must look at Weber's results to see how they build in with these considerations and what the results are."

---

3. In "The Correlation and Conservation of Forces" by E. L. Youmans, New York, 1867, p. 376. Further see W. H. Bragg, *Michael Faraday*, 1931, p. 22 and 25. T. H. Gladstone, *Michael Faraday*, (London 1873), p. 123 ff.: "His Method of Working".

4. John Tyndall, *Faraday as a Discoverer*, London 1870, p. 186.

5. 1.1. [It is unclear to what work this citation refers.]

"Astonishing how great the precautions that are needed in these delicate experiments. Patience. Patience."

"Query these results."

"Must clear all this up by further experiments."

"The hypothesis is not so much mine as one renewed from old times. Look at Euler's letters and what he says."

"Let the imagination go, guiding it by judgment and principle, but holding it in and directing it by experiment."

"Consider for a moment how to set about touching this matter by facts and trial."

"To point out or to lead to a knowledge of what it either cannot explain or has not explained, is quite as important for the progress of knowledge as to establish what it can do."

The analysis so far reveals that true, i.e., new Future as believed in by Faraday takes the form of commands. Whereas the grammatical form by which we project past occurrences into the future is the so-called Future tense: "The sun will rise tomorrow at six. Tomorrow, the letter will arrive in London. Your convalescence will require one more week." Faraday's grammar knows the genuine Future which appears in the form of the imperative: "Consider, Query, Patience, Must clear, Let the imagination go." The command differs from the mechanical future. The latter predicts that the past will go on. The imperative prescribes that something new shall interrupt this predictable march of events. The curse of our time is the idea that the specious mechanical future of prediction is [as] equally representative of "Future", as the imperative. Hence, when people analyze the meaning of the term, "future", they analyze the grammatical form of "It will". But the bedrock of genuine future is in the imperatives which we read in Faraday, or for that matter, in any creative life; in these cases an imperative crosses out all the causal processes by a break in continuity. Faraday *is* such a break in continuity. And this is incarnated in the grammar of his diaries. This pure future comes to us as commands and the predicting Future of, "it will rain", is secondary to the imperative "Drop the atomic bomb", "Take this train", "Become a doctor", "Don't go to college". The character of the future as completely different and independent from the past, stands out in these imperatives with primeval distinctness. We may use circumlocution and say: "I shall not go to college, after all". But this then is simply the report to a third person of an inner imperative already obeyed. When Justice Holmes said to the usher who helped him into his coat, "I shall not be back tomorrow", he used the specious future of reflection and report. But to the President he wrote the same day: "I bow to the inevitable". In other words, in his letter of resignation he admitted that he had—it was on the same morning in court—received the clear imperative: "Resign". Without this imperative, neither his famous remark to the usher nor his letter to the President make sense.

With the help of his diaries, another grammatical fog is lifted, this time from the present tense of human language. The present state of mind in Faraday is one of suspense, "It was almost with a feeling of awe that I went to work, for if the hope should prove well founded, how great and mighty and sublime in its hitherto unchangeable character is the force I am trying to deal with, and how large may be the new domain." Or "After all, there is much which renders these expectations or similar ones hopeless"; "Such beautiful delicate indicating curvatures"; "Strange"; "Of a sudden all wrong and I do not see why. I begin to despair".

The normal form of his present is emotional and exclamatory. It is a specious and indirect present which we call the present indicative. The subjunctive is the normal approach to our present state of mind. True enough, the Yankee will not have it so nor will the grammar book. They insist that the circumlocution of "This is beautiful" holds the same rank of truth as Faraday's honest shout, "Such beautiful . . . lines!" The genuine forms of speech in a present are not indicatives but emotional exclamations or affirmations. As Faraday wrote, "How great and mighty and sublime is the force I am trying to deal with". That is man's present, in fear and trembling, if he is not in the grammar school of the logician or in college but face to face with the forces we are asked to deal with, our present is an exclamation and a suspense.

The place for the indicative of scientific grammar is neither in the future nor in the present of a real man like Faraday. But listen to this noble series of indicatives:

"There was a fire on Thursday evening in Broad Court, Anny Lane. The clouds were low and received a strong illumination from the fire beneath them. The angle taken from the top of the Royal Institution by a quadrant formed by the clouds, the Institution, and the fire, was 24 degrees. Hence the height of the clouds will be . . .

equal . . . to. . . . ." Or, "Soon after sunset observed a cloud forming just [over?] the brow of Shakespeare cliff. It streamed inwards, increasing in size, but all seemed to pour nearly from the same spot; the air which came from over the sea there taking on a visible form and passing in to the interior as a cloud. By degrees the generation of clouds took place along the whole line of cliff from Dover to Folkestone hill, the hill still carrying the portion formed over the land. We ascended the cliffs about half a mile beyond Folkestone hill about half an hour after sunset and found all above enveloped in dense, moist mist, so as to deposit water on our clothes; the temperature also low to the feelings."

The real living person of a Faraday meets the Future by commands, the Present by exclamations, the past by narrations. But the unscientific mind confuses all this. Listen to Faraday: "What a weak credulous-incredulous, unbelieving-superstitious, bold-frightened, what a ridiculous world ours is, as far as concerns the mind of man. How full of inconsistencies, contradictions, and absurdities it is. I declare that taking the average of many minds that have recently come before me (and apart from that spirit which God has placed in each) and accepting for a moment that average as a standard, I should far prefer the obedience, affections and instinct of a dog before it."[6]

Faraday himself, by his clear distinction of command, exclamation, narrative, by obeying the three styles of Future, Present, Past, could rise above this dog mentality of the ordinary human mind. This was emphasized by Faraday himself when he wrote: "Electricity is often called wonderful, beautiful. But it is so only in common with the other forces of nature. The beauty of electricity or of any other force is not that the power is mysterious and unexpected but that it is under law, and that the taught intellect can even now govern it largely."

"The human mind is placed above, and not beneath it, and it is by such a point of view that the mental education afforded by science is rendered supereminent in dignity."

Here we have the terms, "supereminent", "dignity", "above" and "beneath", as attributes of the mind. We shall have to explain this ascent to Olympus, this emergence to some "higher" eminence from the dog mentality. And we shall do so when we return to the religion of the book title, The Nature of the Physical World. For the time being,

___

6. Letter to Schönbein, July 25, 1853, ed. By G. W. A. Kahlbaum and F. V. Derbishire, London, 1899.

we have to explain the Olympian mood of the research worker. It results from the divine freedom which he has. The great thing about science is the right to systematic error. This frees them from the consequences of error which hit the ordinary shepherd or sailor who makes a mistake. The shepherd in Montana perishes if he makes one serious mistake about the weather; so does the sailor. The admissible margin of error in the life of ordinary working people is—to speak quite arbitrarily—perhaps 5%. In Faraday's 16,041 experiments, about one per cent were successful and the rest was error.

Why is that so? The scientists have been set free for the purpose of systematic error. Science is a systematic and voluntary relapse of society into all possible errors. The shepherd cannot afford to suspend judgment. Nor can the man in the airplane or you in front of your class, or Mr. Roosevelt on the day of Pearl Harbor. Life's battle is immediate. Faraday in his lab, may err a thousand times with impunity. It takes the complete isolation of a lab to establish the privilege of making innumerable mistakes. We cannot experiment with war; we must win or perish. We cannot experiment with marriage or we shall never know what "for better for worse" means. But mind you; science begins and recurs exactly at the point at which the mistakes do not matter or at least matter much less than in immediate living. When we have made sure that the number of mistakes may be legion, we have cut loose from the routines of habitual manipulation. As long as we can only allow for say 20 or 30% mistakes, we still move under the pressure of life's battle, and therefore cannot be quite objective. We have not moved into the realm of science unless we know that we are free to make countless mistakes. Innumerability is essential to the errors of science! In this poetic realm of suspended judgment the emotions of purgatory are infinite in number. As Kant has said: all research is tumultuary. Scientific doubt is not the doubt between good and evil. Scientific doubt combats thousand and one possible explanations. It is always bad science which thinks of a "black and white" solution. The number of possible solutions must be enhanced above this purely logical level of an either /or before we can speak of scientific research at all.

The physicist's experiments are not reactions to the past but anticipate or pre-empt a future. Faraday's experiments were not experiences because he undertook them in the light of his faith in the unity and infinite definity of Nature. But then, his work done during the forty years of his present-day life, received its sanction and its authorization not from the past but from the future. Science is provoked by society's faith in a free and different future. Science is the pronaos, the vestibule of the future sanctuary of mankind. The laboratory of Michael Faraday is a vestibule in as far as a genuine future which differs in quality from the past, is believed in and finally is incarnated. The scientist in his research is exempt from the Law of Laplace under which nature groans: "We ought then to regard the present state of the universe as the effect of the anterior state and the causation of the one which is to follow".[7] This law of Laplace is not valid for the man of science himself. Faraday's present was not at all caused by the past. Indeed, all the past before he lived contradicted his faith. His vision of nature was not anything of the past. It marched ahead of him. And the spelling of Nature with a capital N always means exactly this; he who spells it this way, reclaims Nature to be a power of the future under which we may gather against the past. The research man gets himself out from under the past.

Mankind's Future logically precedes mankind's present. We have no present as soon as we lose faith in the future. What we call present, is the result of a struggle between the future and the past, in us. Mechanisms are repetitive; science is not or it ceases to be science.

The scientist is the miracle which supersedes the course of nature and interrupts its trends. The physics which the physicists discover are causes and effects which always have existed. The physicists, however, who discover them, have never before existed. Their faith emancipates their present from their past.

And it is not the physicist only who must have this faith. All the Archimedes in Syracuse may be murdered by the soldiers. The laymen, therefore, and the scientists must be steeped in the same faith. You think that the facts found by science are after all for believers and unbelievers. Be not mistaken. Physics itself is impossible among unbelievers. The faith in science is a condition for its existence. And this faith, "There shall be science", is in itself no scientific proposition at all. It is a social imperative of religion. Societies have been and shall be again which reject physics. Our society had to learn that physicists were not witches. And that took a change of faith, with the nations.

7. *Theorie Analytique des Probabilites*, 1902, p. 3.

For the nations, Nature had to become a beacon of faith shining in the darkness of the world before physicists could get away with experiments.

Fortunately, nations do change their religions. It is an old saying that it is easier for a nation to change its religion than for a scholar or scientist to change his mind. The task of the 16th century was to change the nations. We have come on the more difficult time of history at which the scientists must change their categories. They must re-recognize the religion which unites them with all people long before the first experiment in any laboratory can possibly take place.

The founding fathers of any science always live by faith alone. But the people in the amply equipped laboratories do not need the stout heart of the Curies in their garage or of Heinrich Hertz in his barn. And it is the mass of the employed scientists who endanger today the future of science because they ignore the faith which the founding fathers and the community must share before the city of science can be built. The imperative, "let there be science", is pre-scientific. It was spoken over Faraday before he himself could call himself a scientist. Who then is this strange fellow who all of a sudden, in the midst of life, may proclaim that there shall be science?

THE PROGRESS OF PRAYER AND OF SCIENCE

With this strange question, we are back to Eddington, because the Gifford Lectures are meant for exactly this man.[8] He must be neither a scientist nor a fool, neither a man without leisure nor a man without seriousness. If we could find out who this human being is who reads books on the Nature of the universe, or at least is expected to read them, we would have found the true religionist of science, the man inside the scientist and the physicist and within the laity, both, through whom science becomes real.

We are looking for the believer in science. [One] who believes in its processes before there are any results. And I may say in advance that this man must be appealed to by the treble formula which puzzled us

8. *Ed.* The Gifford Lectures were endowed in 1887 by Adam Lord Gifford with a gift to the four Scottish universities: Aberdeen, St. Andrews, Edinburgh, and Glasgow. Their focus was to be natural theology. Many distinguished thinkers in philosophy, theology, and the sciences have been invited to deliver them, to this day.

so much, the formula of quasi magic in which the nature, the physis, and the world resounded all three. Because this man in you and me will not buy the book otherwise.

Of this strange law, an easy test is available, a veritable experimentum crucis. Take the three following book titles: The World, Physics, Nature, and ask yourself what you would expect behind their cover. The title "physics" would make you expect a textbook on physics. The title "Nature" would make you think of Thoreau, Rousseau, or poetry. The book, The World, might be by Wendell Wilkie or Mr. [Karl] Haushofer [1869–1946]. The three books, in any case, are written for three groups of readers clearly, because they cater to quite separate interests.

It follows that "The Nature of the Physical World" must be addressing a fourth man in you and me. He is not the romantic Rousseau-ite in us, nor the practical student of physics, nor the man of the world, the politician. For heaven's sake, who is this man?

Faraday has paved the road for an answer to this question when he said that the human mind is placed above and not beneath the forces of nature, by science. The man of faith wishes to rise, to ascend above his state. The book The Nature of the Physical World is not written for the practical man or the man of leisure or the politician, but for the man of history, the man who by his faith makes history while all the three others, the practical physicist, the politician and the romantic vacationist, are his parasites. It is the man who takes risks because he can experience and bear with both, the being beneath and the rising above. The man who buys the textbook "Physics" buys admission to the standing above without first sharing the darkness of the World. The man who buys the book on Nature, plays with the world. And the mind who takes the world as it is, prefers permanent darkness. But the living man is in process and is able to experience the change from one level of consciousness to another, and back again. The real man can be overwhelmed by wonder and can see nature as brought under law, both.

The complete man is not the layman or the physicist, but the man who is alternately capable of the wonder before and the precision afterwards without ever thinking that one alone is enough. It is the man who towers above his own fission in a mind and a body.

This is implied by our worksheet. Only I must now invite you to read it once more with ultimate precision. I challenge you to reject the notion that the left page was written by the student's mind, and the right page by his body. It was one and the same who wrote both pages. This man did not have a body and did not have a mind as we usually put it. He became all body and all mind, in alternation.

He tried hard to make his mind meet all other minds in his computing. And he tried hard to immerse his body into the cosmic continuum of the material world. He split himself as best he could. But he did not succeed completely. Please reflect on the mystery that the mind right and the body left, both have scribbled and written. It is true that the style of the handwriting on the two sides differs. Nevertheless, in each case, the same hand moved over the paper; the same fingers clutched the pen, the same shoulder turned. His eye looked on. His elbow rested on the table; his buttocks pressed the chair, in both cases. His feet dangled. Also his brain was at work when he registered as a sensorium. And his senses were alive while he computed on the right hand side. The same person used the same faculties when he registered as a body and when he computed as a mind.

But then, an incredible hoax has been perpetrated by those who try to tell us that mind and body are two things, that man should have a healthy mind in a healthy body, etc. Yes, in my judgment it is a hoax. No such two things exist. The mind is me when the sequence is: brain, eyes, hands, fingers, buttocks, skin. And the body is me when the sequence is skin, buttocks, arms, eyes, brain. I do not have a body and I do not have a mind. The same being, in both cases, is arranged differently. The style of the two handwritings proves that on the left side, we have man in slippers so to speak; then he is as much mind as possible and on the right side he is in full battle dress, as much body as he possibly can squeeze out of himself.

When we speak of our mind and our body, we speak of different styles of being. We may be a mind or a body, in alternation. The whole man is present in both. Mind and body are modes of myself. And they are not arbitrary tendencies. They are enactments of my own faith and decisions.

I become mind and I become body because I alternate between the outer cohesion with the material world and the inner fellowship with minds. At any time in history, this hinge between an outer and an inner relationship of myself, has existed. But in the science of the Renaissance, these two opposite styles of being a body and of forming a mind, reached their absolute and ultimate perfection. Not with some accidental mind but with all minds must my mind square in mathematics. And not with some little corner of the globe but with the whole universe must my body vibrate, in physics. The body of the physicist moves in a more complete world than the body of say a swimmer. And the mind of the physicist moves in a more complete mental continuum than the mind of the friend at dinner. The conditions of mathematics and physics must be such that anybody on earth would make the same experience under the same conditions.

The consequences of our discovery that mind and body are styles of being, are considerable. If it is true that I go in to play the mind and go out to play the body, the creation of this very alternation becomes the true and central concern of the living man. To be able to pass from the outside to the inside and back again, becomes the real crux of life. Never can I hope or wish to be all mind or to be all body. The nudist and the philosopher both are undesirables. My foremost attention must be directed towards being the door into both forms of being. My ego may be the mind who gets his name immortal as Ampere or Volt. My It may be my unconscious body. But you, the person, is the threshold and the gates, the ambivalent and ambiguous free creator of the body- and of the mind-situations. As I take the elements of my being into the outer or the inner world, me is both, the Ego of the Mind and the It of the Body and yet more than both together. Me also is the hinge, the either-or, and the this and that, man's soul is gates and a door. And it never is it more than when he institutes the scientific process. The soul is the hinge which allows us to decide over our mental and our physical style and which enables us to be scientific. In order to do justice to the title of Eddington's book, we had to reach this one conclusion. The book was written for people who have a soul which is free to immerse herself as body into the world and as mind into the fellowship of minds.

From 1500 to 1900 the fact that man was a door could not be mentioned without ridicule. And this made all other civilizations and people inaccessible. Modern Western man seemed so different from all other men as they all stress this existence of Gates and doors. And we laughed it off. We need not laugh any longer. The people of antiquity are brothers.

Now, by rediscovering this quality of the soul as the condition for science, we may re-establish our identity with men of other civilizations. The men of these other epochs recognized that man was the master over two spaces, an inner and an outer. All ritual all over the world and all magic testifies to this faith. This seemed bare superstition. But it was superstition solely in as far as their world was limited, with the Egyptian sky or the Peruvian sky making law for their sky-worlds. Our arrangement of all sky worlds into a sky world of sky worlds, the whole physical world, seems better. It is indeed the most radical arrangement, among all the arrangements of two spaces, one of the mind pointing inward, and one of the body pointing outward. Among the many thinkable gateways between an inner mental and an outer, physical space, our natural science is based on the most universal form. It is an optimum solution among its equals. But it still is their equal in this distinction or alternation of an outward and an inward process. The world of the outer space does not any more exist than the world of the mind on which we insist. Existence and Insistence are corollaries of our own arrangement of two spaces. As Faraday lucidly writes: The mind is placed above the outer space when a whole second world is formed by all the minds in society which we delegate to cope with the three dimensional spatial realm as expert physicists.

All societies delegated their priests for the same purpose. The priests of science may be better priests but they are the priests of the people's faith, just the same. One day, "We", the community, decided that a certain group among us would, over centuries, be free to move back and forth between the two spaces in alternation. Whereas the Egyptians assigned the Nile valley to their priests for observation and computation, the Christians believed not in the Nile, or the Yellow river valley in China, or the Gulf of Mexico. They believed in One world. And so the physicists from the new day of science, were under orders not to think of a smaller world than God's complete universe. The physicists of the Renaissance received their marching orders not from any Greek tradition nor from themselves but from the common faith of Christianity in the Oneness of the created world, as one whole, as that "infinite creature", which the Cardinal Nicolaus of Cues had called the world, and which the creator had called into being.[9] While all worlds of the ancients, of the Greeks and the Hindoos the Chinese and the Mexicans, were finite, the World which God created according to the first article of the Nicene Creed was infinite. Infinity in every direction distinguishes the World of which Faraday searched the secrets, from all former so-called "sky-worlds", from all non-Christian natures of the cosmos.

Otherwise, the modern physicist works under the same conditions as the old priesthoods. The outer world of three dimensions, length, width, depth, does not exist except as corollary to an inner world in which all minds unite. This inner world in which the scientists through the last four hundred years of physics have united, has not three dimensions of space at all. Neither has this inner world and twin space, the same time as nature. In the outer world, time may be considered a fourth dimension of space. We have seen that the predictions for mechanisms make all time one-dimensional so to speak as the whole predicted time span is nothing but the past continued ad infinitum. The time which prevails inside the republic of scientists is of an antithetical nature. Here the present of the physicist is cut off from the past. There can be no science under fate, under a time which is the fourth dimension of space. Scientists live by faith in a future which differs in quality from the past, and create a present which is not dependent on the past, by this faith. The time of the scientific world is composed of three tenses, and the pull of the future and the push from the past combine to produce a present of scientific research. Time is three dimensional in history, in the history of science.

Society expects from the inner space in which all minds may become one mind and from its three dimensional time, new revelations on three dimensional space. The physical space examined by the physicists is only one of the two spaces postulated by the existence of a science of physics. The other space inside which physicists write monographs, keep each other company and communicate their ideas to each other, does not form a part of the space of their objects.

The Science of physics is a historical mandate given to a group of people by Christianity at a certain moment and for a certain future. Eddington himself uses language created by this mandate: "The

---

9. Nicolas of Cusa (1401–1464), a German churchman of many accomplishments in philosophy, theology, and proto-science.

physicist is accustomed to state lengths to a great number of signifi-cant figures. . . . These lengths *are a gateway* through which knowl-edge of the world around us is sought. . . . *The first step through this gateway* takes us to the geometry *obeyed* by these lengths. . . ."[10] He has no other language at his disposal except the commands of religion: *A Gateway has to be built.* It is a cheap escape to call these commands metaphors. They are *indispensable* metaphors.

And the faith of the laity in this gateway is just as much a condi-tion of successful physics as the skill of the experts.

Eddington, in his booktitle, appealed to the fundamental faith of society which called physics into being. We have found the key to his magic formula World Nature Physics.

The key to their explanation is in our hands since we understand the three tenses and the rhythm between the three tenses in the life of science. Each of the three terms connotes one of the three tenses. 1) World, the anglosaxon term, is the world before we really know it. A man goes into the world. And Milton says: The World was all before them. . . . . This world is full of riddles, full of powers that be, full of surprises. World Wars, World Crises, World revolutions may remind you that the World to this day still has this quality of being unpredict-able and of being not our home. I am not a match for this world; no man is. This world puts me on the run. 2) Physics, Physis. Physical—these Greek terms are used when we have succeeded in explaining this same world. This physical universe has become predictable. The world which no longer holds secrets, is the object of physics. We stand above it and overlook or survey it in our courses. We have made it speak to us, in figures. The two terms world and physics are the two tenses of reality before and after we have science.

The man who says Nature is the man to whom the world already is a task of his faith, but not yet a result of his work. He no longer is panicky as the individual who is no match for the world. He has calmed down to asking the question together with others: what are these forces and powers which destroy us when we meet them every one of us, alone? Nature is the question for the universe which men in fellowship have the courage to ask. Singly, nobody has any choice: the world frightens and defeats his mind. The community is the unit in which the monster can be faced and confronted.

10. p. 160f.

He who meditates over the nature of anything, has turned from his flight. The individual is chased by the world and never at rest. It is inexorable that the world keeps us in constant movement. Meditation itself is the act of faith by which we turn around and this is not possible outside the peace of a community.

The results of our confrontation is yet unknown when we say "nature". In clear distinction from the figures of physics, nothing as yet is deciphered. Nothing as yet can be predicted. But the trend of running before the impact of the world, has been stopped, and a counter movement sets in. The man who asks what is the nature of war, is not at war. He has gained time. He is establishing the gates between the two trends of being chased by war and of examining war.

Nature is the threshold word of our language. It describes man's power to turn about towards a part of the chaos around him with the courage to confront it. Nature is the turning point at which we erect the doorway between mere blind experience and impressions and our inner response. This turning point says: so far, every one of us singly, has been made to run. Now, we as a group, pause and look around. The good old term is, we reflect, we look around collectively. Nobody can reflect except as a member of the common peace. The term Nature creates this room for inner reflection. It balances the idiomatic term World of the panicky individual and the learned term physics of all mankind.

This fundamental tripartition of the objects of science according to the three phases of their treatment by us, is valid for all scientific research of the last five hundred years. Take

God Deity	theology	Divinity
husbandry	morals	economy
workers labor	tactics of	labor policy
teller	numerals	arithmetic
healing	medicine	biology
man	humanity	anthropology

I myself have written on "The Revolutions of the Christian World". In this case I spoke in precise parallel to Eddington. "Revolutions" corresponds to "Nature", "Christian" is Greek corresponding to physical, World is identical.

| The Word of God | Scripture | Biblical Criticism |

The same Greek root may serve different sciences, but their different Anglosaxon and Latin predecessors then prove their profound inner difference. Compare psychology and psychoanalysis. They are confused often because both speak of the psyche. Go back to their Anglosaxon and Latin phases, and they deal with a different topic. Psychology is preceded by the soul and the person. The naive individual believed in the soul, the Person was the communal and social question, the psychologist discards the soul just as Bertrand Russell discards the world.[11] But psychoanalysis deals with the Anglosaxon sinner and the Latin Ego. The sequence sinner, Ego, psychoanalysis is proven by the psychoanalytic patient who does repress something. I therefore, find the two trebles, soul, person, psychology as against sinner, ego, psychoanalysis, especially illuminating. Another triad is people, society, masses; here mass has set its name to a science but is in back of the hybrid, sociology. Even popular science still obeys this law. Eddington could have written a so-called popular book on the secrets of the universe. But his publisher would have sold the book by putting on the blurb: by the Nobel prize winner and great physicist. In other words: Even the camouflaging of our law does not abolish it. The book still sells because the author has the Greek name. Miss [Margaret] Mead may write, "And keep your powder dry".[12] But she sells her wisdom on man's humanity as a learned anthropologist. Behind the most revolting race for a catchy title the solid faith of society in its scientists shows.

Our observation of three phases places the scientific process in the historical realm of three dimensional time, with a future which is free from the past, and a present created by faith.

This gives the first explanation for the usage of beneath and above, higher, superior. Nobody has ever tried to show how this steppingstone from below to above is established. Because nobody has paid attention to the necessity for an about face, from mankind to anthropology, from fire to pyrotechnics. Take the music of Wagner

11. *Ed.* Bertrand Russell (1872–1970) 20th century English philosopher with a popular following.

12. *Ed.* Margaret Mead (1901–1978), author of *And Keep Your Powder Dry: An Anthropologist Looks at America* (1942)

on the fire around Brunhild. Poetry tries to reproduce the wild fire under whose impression we are awed. Science, on the other hand, is pyrotechnics looking down on fire and manipulating or managing it. Our faith in the arts and the sciences accepts both states of mind as corollary. One produces the other incessantly, or life dies. The weakness of Eddington's book, by the way, is that he does not understand the interaction at all. He has a static and logical conception of the two states of mind. Art and Science condition each other; he ignores this.

The world is before us; nature is with us; physics lie behind us. And who is this "us"? It is the eternal creature man who in any moment of history must be capable of being awed by the wild, of facing about for the crusade, and of delegating work to the experts. If we wish to live at all, we must allow for the perpetual interaction of all three tenses. The next science under this law may be a science of wars. He would be soulless indeed who could not say: "O world war, world war One, world war Two, and now for heaven's sake world war Three, o destruction, o atomic bombs, let us not go on with them. Come to our rescue, nature of war itself. Turn around"! What is the nature of war but the parricide in all of us, the same belligerency which made me fight through this paper. Belligerency, conflict, parricide, therefore, hitherto running away with us, should re-align with us and become our tool in fighting war by a science of polemics. Yes, we are parricides. War, parricide, polemics, may well be the next triad in the march of progress. If so, it would only happen by no longer shunning war but by facing our own belligerency. The League of Nations and the UNO are silly because they exorcise war as war without ever stopping and making a full about face.[13] They deny their own belligerency and call themselves peace-loving. They are therefore, utterly pre-scientific and religious. Theirs is not the triad of progress but the superstition of panic. They treat their own nature not as polemical. And so a third conflagration is bound to occur. He who does not aver that Cain is in his heart, can never rise to the occasion of creating peace. Science does not evolve naturally. Science is an unnatural rise to the occasion.

And the unanimous voice of history is on our side. The ancients knew of the threshold value of contemplation. And I will now risk any good impression I may have made so far as a thinker by introducing the pagan prayer of Greek origin which in 400 B.C. and for seven

13. *Ed.* UNO, somewhat outdated abbreviation for the United Nations.

hundred more years was prayed annually for the pacification of the township of Rome. In it, the Arval fratres prayed for defense against pest, plague, dearth, to the God of Mors, death, to Mars.[14] You need not remind me that the gap between their prayers and modern science is profound. Granted that it is, one point, the central point which modern man must recover, [that] they and we have in common: they knew and practised the about face as a group, the very step which our routine scientists, routine bigots, and routine politicians abhor. By this about face, their faith created a gateway into a free and better future, just as Faraday's faith created his 16,000 experiments. Do not despise to look at the similarity. Our whole college education after this war will go stale, if we do not confess our humanity as a group which must turn about.

The text of the prayer is simple. Every element is repeated thrice. The center is held by the abrupt verse in which the God is invited to turn. He, Mars, so far leads the attack of all the evil powers against the city's bounds.

Now he is implored: Leap upon our threshold, stand there firmly. This re-alignment accomplished, the evils become blessed elements of welfare. And now the god is for them whom he slew before.

An ominous and sinister power outside their ken, greater than they[,] the Lord of Death, is conjured. By naming it and by analyzing it, they feel that already they have to some extent lined it up on their own side. The same Mars who a moment before bore down on their fields as the wilding—*ferus* is the word for the wild beasts—now has given them some of his own ferocity. This is the meaning of their song of triumph. On the other hand, in this very act, he has turned his direction and looks from their threshold outward while he before drove inward against them. Death becomes Mars and Mars becomes triumph! In Aischylos [Aeschylus], Seven against Thebes, 705 ff., you find a telling parallel.[15]

Not by accompanying the trends but by turning in a courageous fellowship, do we rise to the occasion. We create a change in the world if we dare to stop and to rename one of its elements as part of our own nature. War drives us panicky. Belligerency is an element of

14. *Ed.* Mors was the personification of death in ancient Rome, sometimes associated with the god of war, Mars.

15. *Ed.* The page reference is unspecified.

life which may be put to good use or bad. It is indifferent. The Latin term which the group fixes on a part of the world, has this quality of making it indifferent and thereby making us free to manipulate it. It always takes a change of mind to establish such a threshold. Faith in a future freed from panic, has the power to build such gates. The hinges in which the door of "Nature" swings and by which any part of experience may become manageable, is our own speech. We speak to each other where before everybody had shouted for himself. And the world quiets down and licks our hands. But this right word of ours is not found without an excited social upheaval. The right word is not a logical deduction, but an act of faith in our sharing some quality of the monster ourselves. This partial identification with the world in terms like "nature", with God in "deity", with material interest groups in "labor", with war in "parricide", with sinning in the inflated Ego of us all, is the bold moral act which is at the bottom of science. The scientists must tell their students that science stems from faith.

This faith is not a private but a vast public and historical experience. And it now is possible to answer the most fundamental objection which usually is raised when such strict laws of speech are discussed as we here have discovered. People say: It is impossible that the vernacular, the Latin, and the Greek play intellectual roles. Speech is too accidental and arbitrary. It does not help, in such a case, to point to the Dictionary which on every page bears out our contention. Words have lost their meaning, speech has lost its creative significance for the modern mind. The forty thousand words in the Webster carry no weight against a conviction that words and usages are arbitrary, must be arbitrary. For the free thinkers religion depends on this dogma. The individual words are traced to their etymological origin. But whole layers of inspiration are not discerned. In a microcoscopical example, I shall try to expose the shortcomings of this attitude first. And from there, I shall proceed to delineate the hourglass which has been created for the perpetual translation from the vernacular into Greek via Latin.

The example which is intended to prove that the modern mind is dogmatically prejudiced, since the Renaissance, against the functional interaction between the vernacular, the Greek, and the Latin, is a mistake in translation made by Luther and the King James' Bible. In John 19, 20, we read of Pontius Pilate's inscription on the Cross INRI,

Jesus Nazarenus Rex Judaeorum. This formula INRI is expressed in the language of Rome. Accordingly, the gospelwriter adds the following remark: "And it was written in *Hebrew, Roman, Greek*." The Jews had resisted Greek influence. Rome which brought not philosophy but the sword, forced Jews and Greeks together into one world. Peter went to Rome from Jerusalem. And Luke wrote his two books so that the Lord in Jerusalem and Paul in Rome might be shown in parallel. The Roman language of INRI is the form of the inscription which is quoted to this day. And the gospel speaks not of Latin but of Rome's tongue because the Roman Empire spoke, not some Italian landscape. However, our translators in their Renaissance mood, changed the unanimous tradition of the text. Luther and the Authorized version changed the order of the original: "Hebrew, Roman, Greek," and instead they wrote without any basis in any manuscript: "And it was written in Hebrew and Greek and Latin." Not the living relation between Hebrew, Rome, Greece, at the moment of the crucifixion but the classroom knowledge that we put the Hebrew first, the Greek second, the Latin last, dictated this translation. A scholastic sequence of languages to be learned, took the place of the vivid picture of an interaction between the Jews of Herod, the Romans of Tiberius, the Greek of the traders and rhetors.

After this, it will be understood that the Church created on hourglass between the vernacular and Greek, *id est*, the language of the mind and science. The Church interceded between the language of the mind, for a thousand years. As the heiress of Rome, the Church spoke Roman.

Our translations of the text of John which do not say "Roman" but "Latin", while John wrote "Romaisti", weaken the significance of the act of Pilate when he wrote INRI. So do we when we call the whole layer of terms which stem from the language of the Roman Church, merely words of Latin origin. It was unimportant that Nature Person Society Ego, morals divinity, were words of an Italian idiom. It was relevant that these terms were parts of the language of the Church. For it was thereby the language of the place in which the Gentiles learned to face the most heinous and hideous features of their own panic. The folly of folks who whirl in isolation, is unlimited. The Church was the meeting ground on which the Gentiles learned to face-about, to turn upon themselves and to form a fellowship which could cease to shout and could face God's Person, the World's nature, and Man's societies. The Renaissance of Nature's Science was preceded by a renovation of the science of God's Persons, of theology[,] and it now should be followed by a science of society's conflicts. But this sequence of sciences from Anselm to Freud makes sense only because the language of the Roman Church gave the laity the courage to put all these awful and awe-inspiring issues on the agenda, one after the other.

The process resembles an hourglass, with the Roman of the Church forming the small aperture between the vernacular and the Greek. Everybody knows of this hourglass in practice and testifies to his knowledge by speaking of laymen and experts, of laymen and clergy. Our whole discussion has simply gone behind this usage of the term "layman", laity, and given him the linguistic status of the man who speaks the vernacular, who does not yet speak the Roman of the Church nor the Greek of science. But consider the astounding fact that the laity has two opposites: one contrast is formed by the clergy. The other contrast is formed by the scientific experts. The layman has, in other words, two groups which work on him: the Roman and the Greek, religion and science. Both came upon the laity with the dignity of something ecumenic and universal. Thus the hourglass was construed from which the sciences could proceed and progress. That science is a child of the Christian era, is written into its constitution by the very terms "science" as well as "laity." For science itself is a Roman term and it was under this term that layman could be educated to honor it. And Laity is a Greek term and it was under this term that the schoolmen of theology and the academicians of physics could be made to serve the people instead of using their knowledge for witchcraft and magic.

The hourglass is threatened by masses who hate to be called laymen, by [~~priests~~] [?] who hate to be called [~~ministers~~] [?] and by scientists who hate to be called believers.[16] And this threat stops the process of scientific growth. For, the destruction of the hourglass would be the end of all science.

Wherever we have not yet faced about and admitted our own true nature, we still face destruction. With physics far advanced, we have difficulties in realizing that its birth occurred in the same

---

16. *Ed.* Rosenstock-Huessy, or some editor of the typescript, deleted "priests" and "ministers" and penciled in instead substitute terms that are indecipherable.

emphatic manner, by a jump of the whole man, body and soul, outside the pressure of the world as it then was and looked. Madness, wars, degeneration around us still wait for their physics. And we must pray that the staffs of the older sciences will help us to rebuild the moral fiber and the religious intensity which once gave rise to physics.

The progress of science depends not on the frantic talk about the atomic bomb but on the progress of *rational prayer*. Before we do not face the nature of war, we misinterpret the lesson embodied in the progress which has led us from the "world" to "nature" and on to "physics". This religious intensity is once more reflected in the rather fundamental triad of the book title "The Nature of the Physical World".

Chorus of the Priests of the Roman Commons

(older than 400 B.C.)
To the Lord Mar or Mars
whose elements are in mildew and fecundity,
ruin and protection
pest and health
terror and taking charge
invasion and defense
ruthlessness and fidelity.
"Mars is the power of doing and averting harm."[17]

I.

Ah our Common's Lares, save us; Ah our Common's Lares save us;

Ah our Common's Lares,
save us.

II.

1. And no pest and ruin, Mar, Mar, overrun more and more people. And no pest and ruin, Mar, Mar, overrun more and more people. And with pest and ruin, Mar, Mar, have done for any more people.

17. W. Warde Fowler, *The Religious Experience of the Roman people* (1933).

2.   Be sated, wilding Mars, bound on our threshold, stand, this
spot, this spot; Be sated, wilding Mars, bound on our threshold,
stand, this spot, this spot. Be sated, wilding Mars, bound on our
threshold, stand, this spot, this spot.

3.   Thy twelve elements our twin groups shall call upon in their responsory. Thy twelve elements our twin groups shall call upon
in their responsory, Thy twelve elements our twin groups shall
call upon in their responsory.

III.

This done, ah Mar, Mors, save us,
This done, ah Mar, Mors, save us,
This done, ah Mar, Mors, save us.
End-chant and Dance:
(Now the God has entered us, we no longer call him but he
speaks:) Triumph, Triumph, Triumph, Triumph, Triumph[18])

18. The sound of Triumph, had not developed a nominative at that time but was the
God's own speech from the lips of his people.

The text and translation have been constructed on the basis of the famous lecture
by Eduard Norden at the Harvard Tercentenary, *Aus Altröemischen Priesterbüechern*,
now *Acta Regiae Societatis Humaniorum Litterarum Lundensis*, XXIX (1939), 107–280.

Six

OUR URBAN GOGGLES

An analysis of the future relation
between the City of God and the cities of men[1]

SUMMARY

People in the city live in a peculiar manner; and this manner is bound
to become their second nature. At times we darkly remember that we
also have a first nature in virtue of which we belong to the City of God.
And in these moments we are apt to put all the blame for our own
misery on the cities of men.

I shall not do so. By building cities, we have given a brilliant expression to some of our noblest faculties. The positive achievement of
the city is foremost in my mind. I invite the reader to a sober assessment of our citified nature. With this civilized or citified nature the
trouble is the same as with any second nature. If it is true that the city
produces a highly specialized pattern of behavior, it also is true that a
man's second nature is not good enough for any man. My life, it may
perhaps be said, is a case study of this revolt against our second-rate
nature. For I was made aware with a shock, at thirteen, of the fact that
the city is merely a second-rate nature.

1. *Ed.* According to Lise van der Molen, *A Guide to the Works of Eugen Rosenstock-
Huessy* (1997), this essay appeared in the journal *Religion in Life* (1948) under the title:
"What Is the City Doing?" It was published in German under the title, "*Was bedeutet
die Stadt?*" in *Zeitwende* (1949) and in Rosenstock-Huessy's *Die Sprache des Menschengeschlechts*, Bd. 1 (1963), 221–234, in English, under the title "Die Grosstadt."

This shock has determined my life in all its later phases. Even when I landed in New York in 1933, its effect was continued in my prayer to land me in America, yes, but not in New York. I grew up in a metropolis of American "tempo", in Berlin, Germany. I was sent to a school to which the court and the bankers sent their sons. My class was worth many millions in dollars and in titles of the peerage of Prussia.

At thirteen, I transferred to a strange school. This, too, was located in the heart of the city. However, I was one of only two day-students. The three-hundred-years-old gymnasium was for boys from the small towns of the province of Brandenburg.[2] Practically, it was "Winesburg, Ohio," in the middle of the Bronx. On the whole, the atmosphere was hostile to a day student; I had to defend myself for being from Berlin. And my dreams of the goodness of the countryside were shattered. But certainly, "Winesburg" by sheer contrast opened my eyes to the second-rate character of the way of life in the metropolis and in "Winesburg" as well. Then and there, I came to know—before I ever heard the term, sociology—that second-rate things like local environment must never contain a man. And all the decisive steps of my life have been attempts to check these second-rate natures in myself or others. I do not think that this is said only in retrospect. At seventeen, when we graduated, my classmates told of their plans which all converged on a locality they already knew. I told them that no real life could be lived that way; that one could write their obituaries already beforehand and that I would not stand for such a predictable life. Thus, it came about that since 1906, I have looked for a way of allowing man's primary nature to breathe. Accordingly I propose to make the following points:

1. What the City was doing to the Christian way of life, was pretty well known in, say, 1800 or 1850. However, in those days it also was known that the countryside did something to this way. The Christian way had to strike a balance between the mores of the countryside and the new ideas from the cities.

2. Today all of America is one majestic City. Industry has removed the barriers between city and country. The whole area is citified.

2. *Ed.* In Germany, gymnasia were secondary schools specializing in Classical education.

3. The new citified humanity, however, does divide its time between a fast and a slow way of life. The speed is realized in the centers of production, in factories, and business sections. The more restful aspects of life are represented by our suburbs.

4. We are confused because neither are the factory districts mere replicas of the old cities like Boston or Baltimore, nor are the suburbs simply the heirs of the old-time villages. The essential contrast between the new equilibrium of factory and suburb and the old equilibrium of cities and villages is often overlooked; hence the new onslaught of the City of Men on the City of God is not noticed.

5. The essential contrast lies in the fact that both, the old village and the old city, believed in their words and ideas. The factory district as well as the suburb of our time act on the assumption that nothing they think or say today may be true tomorrow. They follow the trend. They feel entitled to advertise the bestsellers of tomorrow and next year as well. Both factory and suburb represent a new attitude toward the Word.

6. The Christian belief in incarnation, the universal belief in God's creation, the right use of human reason, all three are destroyed by the new City of Men. And this is not done by accident, but by establishment. The new city can't help doing this.

7. Any new equilibrium of natural forces has always threatened the City of God. But the citizens of the latter usually wait too long before they grasp that the City of Men has taken a new shape. In this article, we shall simply try to grasp the new shape of our eternal partner, of "the world" within our own nature.

THE HEART OF THE TIMES

In 1800 or 1850, the Christian way of life was hampered by two enemies, by superstitions from the back hills and by new philosophies from the cities.

The Christian way of life always fights two enemies at once: the "too slow" of apathy and the "too fast" of mere curiosity. Why must this be so?

Well, the ocean's ebb and tide, and the milky way of stars need no churches. Their life cycles are heartless; their times rest with God. We men need religion because our heart's calendar does not coincide with the astronomical cycle. Astrology is nonsense. Any generation or individual or class or nation has its own calendar which clashes with all the others. Men's times conflict. Unless we build up one body of all men through the times and make God the heart of all our times, we destroy each other. The Christian way of life builds one Body of Christ through the times, with God its heart, and thus overcomes the false times of the fathers and the children. It "turns the hearts of the fathers to their children, and the hearts of the children to their parents." Or we may put it the other way round: the Christian way of life puts heart into our times and thereby creates one Body of Time. Without a heart there can be no living Body of Christ.

This Body always has the same two opponents: (1) the hasty, hurried march of time from one blind change to the other, and (2) the tendency to blind repetition, the apathy of mere routines.

The Christian way of life is opposed to change for the sake of change, and to tradition for the sake of tradition. It thereby obeys the divine Will as it stands revealed in the great calamities and catastrophes. For who can doubt that, for instance, the last two world wars have called back the human race into the universal rhythm from which the pride of nationalism had tried to stray?

Before the industrial revolution, the natural function of the old city made sense. The countryside inclined to be superstitious. Down to the Russian Revolution, the peasants of eastern Europe observed the rites of Isis and Osiris. "Neither the Christian missionaries nor the emperors of Rome had scratched more than the surface of their lives".[3] Superstitions are outmoded ways of life. Rural life preferred such folkways. As a natural check on this one trend of our nature, the city stood for new ideas. Here, new philosophies could arise, new ideas be sown, and change could exert pressure in the form of new fashions, new sensations. Between sensations of a new character and superstitions of an old type, the old Adam in all of us muddled through. We all are one half the rooted plant and one half the roving animal; for us, the village stressed the vegetative rhythm of the recurrent seasons, the cities procured the acceleration of changes.

3. James George Frazer, *The Golden Bough* (1922).

We, however, have abolished this time-honored division of labor. We no longer have peasants. In a mutual embrace, country and city have engendered the industrialized world of factories and suburbs. From the remote corners of the countryside, the raw materials which the machines transform are taken; the scientific process by which they are exploited hail from the city. On the other hand, the rhythm of the suburbs seems similar to that of the countryside, but the mind of the people in the suburbs are all trained in the most modern ways of production.

Hence it is not true that our factory districts are identical with the old cities; for this, they are far too close to nature. Neither are the suburbs simply the heirs of the villagers; the people of Scarsdale are too close to Manhattan; who could be more sophisticated?

One similarity between the routines in the old peasant homes and the homes of our suburban commuters cannot be denied. It consists in a distinctly more relaxed, more leisurely approach to the time schedule than either the old city or the people in the Loop can afford. But both, the suburb and the factory, have some new relation to human language which was unknown to either the old peasant or the old citizen of Boston.

The peasant was superstitious in that he repeated the sacred words of the past forever and forever. And when I go to our own village church, one out of three in town, with from fifteen to twenty others, I am superstitious, that is, hanging on against hope.[4] For, this handful of people certainly is not the salt of the earth or the undivided Church of Christ in our town. But there is nothing wrong with the service which we observe. Our words are not superstitious. The situation is outmoded; that's all.

Now, however, turn to the suburb. At the outset, in the new suburb provision will be made for all the denominations—Catholic, Jewish, Protestant, minor sects. No one faith is absolute in claims or expectations. Faiths, in the plural, are a Sunday affair. The suburb is redundant with private activities all of which are perfectly harmless and without consequence. The best book for the suburb is *Alice in Wonderland*. The doggerel is its most pertinent poetry. Dante is funny in the suburb because in the suburb nobody can be exiled for his ideals. In fact, everybody has ideals there and they all differ. People read

4. Rosenstock-Huessy is referring to his home in Norwich, Vermont.

voraciously in the suburb. But in the old village, they had only one single book through the centuries. Hence the villagers would actually believe in what the book said. But the suburban reads the review of a new book before he commits himself. The words preached and read and rhymed in the suburb, all are uttered tentatively and in good spirits. By good spirits, we mean without giving offense to anybody. And that is a good way of saying, without any effect on anybody. For the man who is never misunderstood to the point of offending can never have said anything important. Important words always give offense. They make a difference. The Holy Spirit is not a "good Spirit" but the better Spirit!

Now compare the old city and the new Inner Sanctum of Simon and Schuster. The old city gave birth to philosophies like Spinoza's or Schopenhauer's. Their newfangled ideas disturbed the peace. The idea required partisanship, decision, commitment. Because these ideas created a whole movement, like transcendentalism, ideas made martyrs. Mind you, many of those new ideas were cockeyed and merely new. I do not think that in themselves they were better than ours. I do contend that our ancestors stood by them in a very different manner. The wicked new ideas of the city were persecuted and they were introduced by people who believed in the importance of making a grave decision.

This relation of the writers and publishers to their own ideas is impossible when you write advertising copy, or editorials for a paper whose political convictions you do not share. If a Gallup poll can offer the publishers and authors a poll of what will sell, the last camouflage is dropped. Nobody any longer pretends that he is in conscience bound to write as he writes. He eagerly admits that he is going to write what pays.

The most striking difference between the old and the new relation to the Word deserves to become the theme of a book. The title I planned for it was "The Triumph of the Witches". I wanted to show that the same type of people who formerly were burned as sorcerers soon may run our society in the form of psychologists and economists and sociologists, and may put everybody who speaks only out of conviction into their carefully padded lunatic asylums. The modern mind declares anybody who keeps from writing for money to be a fanatic or "nuts." An athlete and brilliant college graduate who had joined the old CCC in order to reform it, volunteered after Lend-Lease for the Marines.[5] He was rejected by the army psychologist as a lunatic simply because no "normal" fellow could go from college to the CCC. If he had followed the next trend, that would have been sane, even if it had consisted in ruining his health by cocktails and venery.

The new majestic City America, in other words, has developed a new attitude toward the new ideas and the sacred traditions of the race. Everybody is noncommittal. A marriage consecrated by the Cardinal of Boston ended in divorce a few years later. From the Inner Sanctum of a publisher, we may expect every year another creed and another philosophy and another policy.

Words have lost their meaning. Names have lost their appeal. The publishers instead of consulting the Gallup poll should ask themselves if books did not depend for their very existence during the last four hundred years on some strange identity of the speaker and the words he spoke, and whether probably the time for books is over as this identity is lost.

If and since we all ride the wave of perpetual future change, no one single change can ask for our devotion or investment. The business district always has its tongue in its cheek. And in the suburb, we can't ever get excited as this would make us unwelcome at the country club. (The other day I read of a Country Club Church!) And now let me give three examples and then be silent. In these three examples the new City of Man has altered our relation to Christ the Word, to God the Creator, to man, the image of God.

THE PERMANENT WAVE OF THE FUTURE

In the January issue of the *Reader's Digest*, Anne Morrow Lindbergh gave a write-up of her most unforgettable character. Speaking of his death, she said, "The flesh had become word." The author of *The Song of Bernadette*, Franz Werfel, a man whom you might suspect to have religious insight, printed in his last book, "At the end, we shall say that

---

5. *Ed.* CCC was the Civilian Conservation Corps, a quasi-military Depression relief program for unemployed youth. For Rosenstock-Huessy's involvement, see Jack Preiss, *Camp William James* (1978). The Lend Lease Act of 1941 provided material support by the U. S. to the U. K. in its battle against Nazi Germany. It was an interim step before a declaration of war.

we have created God." [Thomas Henry] Huxley and the evolutionists explain the so-called higher by the lower, man by hydrogen, and God by stomach ulcers.

Let us take the undaunted heroine of the wave of the future first.

Mrs. Lindbergh's sentence, "The flesh had become word," rivals the sentence from John: The Word has become flesh. Obviously where people clothe their beloved for the burial themselves, or where the picture of the Crucified is still looked upon in faith, such nonsense would be unprintable. The corpse gives off a stench. This, in the suburb, is hidden. So, the five words, "The flesh has become word," did not arouse indignation. That it was blasphemy was not felt. This brings out the fact that the modern city denies the very possibility of blasphemy.

The modern city does not rest until the last sentence of our faith has been matched by a brilliant worldly parallel. This is achieved by changing the direction of the faithful statement. By the change of the direction, it becomes witty. In "The Word has become flesh," the spirit of God descends. In "The flesh had become word," the human mind is distilled from the body and ascends. Similarly in "God created man," Moses looked in one direction, and in "We have created God," Werfel looked in exactly the opposite one. In the sentence, "In the image of God created he him," all the things below men, oceans and stars, mountains and valleys, are later than God's vision of man. They lead up to him. But with Huxley, the earlier explains the later, the mountains and the molecules evolve man in their image.

All city wit, however, depends for its remarks on the existence of the treasures of faith. Frank Lloyd Wright's son could not have written his biography *My Father Who Is On Earth* without stealing from the Lord's Prayer. Neither Mrs. Lindbergh nor Werfel nor Huxley could have said what they said unless the reverse had been believed by all men for thousands of years.

We discover: the perpetual waves of the future are of a secondary nature. They exploit the treasures of the universal faith of mankind. It took 5,000 years before St. John could exclaim, "The Word has become flesh." It took 3,500 years before Moses could joyfully shout, "In the image of God, he created Man." It took 7,000 years before Niels Bohr could explain the constellation in one atom by the order of the solar system or before Joseph Wittig could explain each individual soul as the replica of the whole church in all its offices and branches.[6] The statements of faith always take time. The exploitation of such gold mines of truth by the city wit takes next to no time.

As we have blown up the forests of millions of years in our steam locomotives within one century, and as we are exploiting the oil deposits of endless periods of geology within this quarter of a century, so the city explodes the accumulated wealth of millenniums of common faith for one magazine article. I am doing it myself at this moment. We all live in this city where the clever mind mints the gold bars of eternal truth into cash.

However, we are now in a position to define with precision the laws under which the City operates.

1. The City exploits the oil wells, the coal mines, the treasures of faith by a change of direction. Lower explains higher, the flesh ascends into the word, my maker is said to be my makeshift.

2. The operation of the brilliant mind seems to be nothing but the act of one day. This is not so. Two ranges of time, one excessively long, one excessively short, are brought together in the operation.

3. The perverted citified statement always remains indebted to the sentence of faith which it perverts, for its creative substance.

That there is a "Higher" in this arbitrary and chaotic universe, that there is a "Creator," and that there is one phase for the word and another phase for the flesh, these substantial truths had lived and had been believed before the direction could be turned about. But of this third law, I would like to say one more word before leaving it to the reader how he is going to restore within his own accounting the balance between the City of Men and the City of God.

May I be pedantic and simply print the sentences side by side:

The Word has become flesh./	The flesh has become word.
Man is in the image of God./	The lower evolves the higher.
God created man./	Man shall have created God.

6. *Ed.* Joseph Wittig (1879–1949), a reform-minded Catholic priest, was a close friend and colloborator of Rosenstock-Huessy's and one of the editors of the journal *Die Kreatur.*

The word which comes out of Mr. Smith's flesh may be anything—a joke or an abomination, a blessing or a curse; there are innumerable unforgettable characters. The sentence on the right side is pluralistic. The sentence on the left side is singular; it has happened once forever, and if it is true, we all live in this One Word's Christian Era; if it is not true, there is no hope for peace whatever.

The God whom men are going to create according to the poor fool Werfel may be a monstrosity, asking for the slaying of our firstborn. The God of righteousness and mercy, however, although he cannot prevent the city people from destroying themselves within three or four generations, keeps the human race alive. "The lower evolves the higher" is a naive theft of the term "high" from the left side of our account. In pure evolution, the word "high" does not exist. The ape is later or more complex than the jellyfish; he is in no way higher. "High" does not come in except by a comparison between God and his angels and men and stones, from the peak downwards.

Whenever the human mind has achieved this perversion of direction, it feels safe. From the corner where the lower explains the higher, where the flesh becomes the word, where we create God, no orders have to be feared for our free will. Sentences like those of Werfel, Lindbergh, Huxley, dissolve our dependence on some imperative truth. For truth is valid only when the singular of a unique demand here and now is heard by the "cross-over" which you and nobody else in the world embodies; if you receive the word into your flesh you admit that the higher overrules the lower and that the image of God may be impressed on the physically ugly, the mentally fearful, the socially underprivileged because it never, never, never shall evolve from the bottom up but always shall descend from the top down.

The little churches today in our suburbs often form part of the evolutionary city of men. The innocent young man in my church one day received new members of the congregation. He had us sing the grand hymn: "The Church's one foundation is Jesus Christ our Lord." And then, with his eternal smile of unruffled suburban kindness, he continued: "Today, we found the Church." He did not even take notice of this change of direction and everybody in the congregation was far too polite to do so.

The City of America does in a new and peculiar manner that which the cities of men always have done. This minister made the same mistake which mars the three analyzed quotations. The reader may catch himself in this act each time he replaces the word "a" by "the," or the plural "men" by the singular "Man." As this is a kind of master key to the worldly mind's operations, I recommend this observation. It's a lie detector. Werfel's formula that man creates God, is false because the tragedy of men is that they can never hope to become MAN except by the grace of God. God must have given us a chance to form ONE SINGLE MAN before we may reveal God. *The City of Man* was the attractive title of a book ten years ago. It was written by the leading liberals. The fallacy was in the naive use of the singular *Man*. With old Homer, it still was notorious that there were "many cities of men"; in this honest manner the *Odyssey* begins. Our liberals jump to the conclusion that we can build a city of Man without God blessing our work. In the same manner, our young minister might have preached on the humble endeavor to found today "one" church, in the image of God's foundation. But he jumped to the liberal conclusion that "the" Church was man-made. And it is not obvious that when Mrs. Lindbergh's hero died, not "the" Word had become flesh but some word, one word among many had been added to the confusion of tongues?

Whenever something indefinite, the "any" or the "a," is exalted into the One by mere cerebration, without personal commitment and sacrifice, it always betrays the humanistic mentality. In this act, the world takes the place of God. We daily commit this act. The great Pope Gregory VII fought this surrender; he called it simony. Luther fought it; he called it indulgences. Julien Benda [1867–1956] fought it; he called it "*La Trahison des Clercs.*" The city of God which fights it will live to the thousandth generation; and the city of men which does not fight it, will have vanished before the fourth generation.

When the minds cease from this mental fight, our bodies get involved in wars, our property in economic crises, our souls in sadistic racial hatreds.

But will anybody fight? Is there anybody left who can fight? The reader who has followed us thus far has a right to say that the new city is omnipotent and therefore cannot be held in check by any Christian way of life. Indeed the City of Man in our time is so formidable because it does include the peasant and the philosopher, between which the old Christian could find his way. The new city dweller is a fusion of both these extremes. This city dweller is repetitive like the old peasant and he has brilliant ideas like the former philosophers. The result is that he is a man who repeats sensations. While in former centuries the peasant used to repeat ancient lore and the philosopher created new ideas, the modern city dweller incessantly has one sensation succeed the other in stereotyped repetition. He has the superstition of believing in a breathless chain of daily news. Every single one of them differs; however, they are repetitive as they are all crazes without consequences. And it cannot be said that waves of the future in endless succession are more intelligent than the endless turning of the prayer mills in a Hindu village.

To fight this new "superstition of enlightenment," no army exists. Our ministers are numbed by this new alignment of forces. They have not "studied" this situation.

The one man who saw this unholy alliance of speed and superstition early is Friedrich Nietzsche. He mourned the death of a living faith. In his despair, he mixed a drink for the dead souls of our peasant-philosophers. His phial contains a counter-elixir, an antidote against this obsession with sensations in succession. Nietzsche volunteered for the only role which can impress such a city dweller because it is the extreme role of this same city dweller's existence. Nietzsche undertook to play the Antichrist. Nietzsche's Zarathustra does professionally that which Simon and Schuster and Mrs. Anne Morrow Lindbergh do only occasionally: he replaces every act or scene from the New Testament with one of Zarathustra's vintage. Nietzsche made himself into the antichrist to resuscitate in the poor breathless souls the power to distinguish the spirits, that is to distinguish between panting and breathing again. He took the devils dress lest God remain dead. We have this from himself. This poem suffices to prove that he knew what he was doing and that we do him the greatest honor if we accept him as the antichrist; Antichrist is an Ersatz Christ, and the city's way of life is Ersatz.

The mind of the city has reached its insuperable absolute in Nietzsche. And against this foil the cross leaps forward with renewed vigor. The city annihilates all ways of fruitful incarnation. Nietzsche replaces Christ. And behold, never is Christ more redblooded and interesting than after you have tried Nietzsche. The Antichrist can stem the waves of the future to which our ministers and Christian fronts and peasant-philosophers succumb. By outdoing all city wits, Nietzsche has staked out the ultimate. The last word of the city: Nietzsche has said it long before anybody who may come in the future. I stand not alone in this belief. But I did not know how literal my agreement with others was on this point. Indeed, this article was sent to the editor before I found the comrade in arms, Gerhard Brom, in the *Nederlandsch Royal Academy of Amsterdam, Transactions of 1946.* He says that Nietzsche's Antichrist has reduced the New City of Man ad absurdum. "Christ walking among man's children, is the Word which has become flesh. But Zarathustra is the flesh which has become Word. This is a parody. It is the weapon of the powerless who wants to make himself big and who remains literature. . . ."

A succession of sensations still is a succession of mere sensations for every moment. And the Christian way of life still is and will be a succession of apostles to each generation.

APPENDIX

Since this has been written an important new example has been added to illustrate the wasteful and exploitive character of modern poetry and fiction.

My friend [Karl] Zuckmayer has staged the French German enmity of the last war under the Biblical title *The Men in the Fiery Furnace.* And when in this play a score of poor French devils meet their atrocious death he has a chorus intone the Biblical song of the three men in the fiery furnace. Now being a playwright he had to do it within the laws of his trade. I know from himself that he did not notice the change that he wrought in the Biblical text. He is certainly no cynic. Hence, the laws of the profession may be studied in this case without any moral bias. This is not saying a word for or against the play or the Bible. However, I do want to show the abyss between the quite unliterary, even antiliterary Bible and modern city literature. For the difference explicitly is denied by most modern higher critics, experts, philologists, and ministers.

The Biblical text runs:

> All works of the Lord praise the Lord
> Laud and exalt Him through the generations
> Praise, ye angels of the Lord
> Praise all the waters which are above the heavens
> Praise sun and moon
> Praise, stars of the sky
> Praise rain and dew
> Praise, fire and heat
> Praise the land
> Praise mountains and hills
> Praise whales and fishes
> Praise beasts and cattle
> Praise ye sons of men
> Praise Israel
> Praise the priests of the Lord
> Praise the servants of the Lord
> Praise ye spirits and souls
> Praise the saints
> Praise Hananiah, Azariah, Mishael. Amen.

I have omitted a number of links in between because I wish to stress that this text has a miraculous order. For, in the midst of the furnace Hananiah, Azariah, Mishael try to keep alive. And they sing the praises of God; they first look up to God's throne and see the angels. They see after that the high heavens; that is to say: in their ecstasy above their agony, the highest and farthest has drawn their attention first. Gradually, however, their power of conscious sobriety increases. The earth comes into their sight, the human race, Israel, the priest in Israel, the saints in Israel, the hearty ones in Israel and at this moment the rope snaps and they dare rest on their own existence, now verified in the light of all higher orders. Sweetly these singing adorants Hananiah, Azariah, Mishael say, with angelic smiles to themselves, to each other, "Hananiah, Azariah, Mishael praise ye the Lord." It is the triumph of their psalm that they finally have the power to say this. Anyone in terrible pain projects this as far away from his self as his thought will carry him: that the angels were invoked first, was natural, but that Hananiah, Azariah, Mishael are asked and requested last, was sublime. The most profound law of analysis, the law of projection, here it is at work in the sequence of these lines.

Zuckmayer's play brings in, as an epilogue, the song at full length and the whole text is given, but the names Hananiah, Azariah, Mishael are omitted. With this new arrangement, the soul's original reason for the whole order of the various summons becomes undiscoverable. The structure now is accidental. Now, the praises seem to be put in an artistic, or rhetorical, or a logical order: men-willed, men-thought, men-ruled. But in the real fiery furnace when the prisoners first drafted the highest angels for the praise of God, they already aimed at the victory of the three singers themselves as it is finally made explicit in the sweet self-address. And vice versa, the final self-address is equal in power to and is of the same high pitch tension of the first line. The angels and the poor Hananiah, Azariah, Mishael then must not be considered as some logical positions x or y. To the contrary, they are the entrance and the exit; and more than that: they provoke each other and each is, in the very strict sense of the term, the *cause* of the other being called out at all. These poor people could not have begun with themselves, but they were only justified in calling upon the angels because they persevered until they themselves felt as free to sing as the angels. Praise, ye angels of the Lord. . . . Praise Hananiah, Azariah, Mishael, is one cadenza!

Zuckmayer then has universalized, generalized the Biblical song as all humanists and has deprived it of its empirical, direct and unique features. Yet in his trade that was or is expected from him.

But then the garbling up of the precious stones and pieces of brocade in the Bible for poetical perusal is responsible for the fact that the Bible is treated as literature and that the cost of truth is underrated. Plays may be written every year. The song of the men in the fiery furnace is one and one only in eternity.

$$Seven$$

LITURGICAL THINKING

Reprinted from *Orate Fratres*, Collegeville, Minn.,
November, 1949, and January, 1950.

In Memoriam, Joseph Wittig, August 22, 1949

I.

The liturgical movement is intimately connected with an upheaval against modern thought. It reflects this change and is embedded in it. How else could it be? Man, healthy man, as he is called into life as the image of the one and indivisible Trinity, cannot move in any one field without moving at the same time in all others. If our mode of prayer changes, our modes of thinking cannot help changing also.

It seems to me that a wider look around may help us to understand the liturgical movement within a larger context. Modern secular man begins to doubt the fruitfulness of the modern mind's logic, science, method of analysis. Perhaps the liturgy itself has revealed and represents a truer way of dealing with life reasonably and truthfully. I therefore have called this essay "Liturgical Thinking." May we perhaps learn from the liturgy how to think on all problems of the mind?

I am inclined to think so. In six decades, I have been led to slough off the standard procedures of so-called scientific logic as harmful. The modern mind of the Renaissance is obsolete. The era of the Reformation and of the Counter Reformation has made too many concessions to this Renaissance mentality. The liturgical movement of the last decades already has eliminated many plaster casts, accretions and trimmings by which, after the Church was rent by the Reformers, both parties of religion tried to reconcile the Renaissance mind to the liturgical tradition. That this elimination is hailed, goes to show that we no longer need to make the concessions deemed necessary after 1500.

Hence, I shall proceed in the following manner. I shall single out, in a first article, some outstanding features of Renaissance, Reformation and Counter Reformation. In a second, I would like to tell what I have learned from the liturgy for a revolution of my own thinking.

PERSON AND COMMUNITY

"Postmodern" man differs widely from the men of the Renaissance. We are analyzed as bundles of nerves. Schizophrenia is rampant. We are torn and often we break down. In 1500, every layman claimed to be a "person." Before, "person" in canon law meant a dignitary, a bishop perhaps, or an abbot, or a princely person. Persons had status and authority. They had something to say, to administer, to answer for. A person was always responsible for a functioning part of the whole community, he held an office of some kind. The smallest "office holders" were the fathers and mothers who presided over households. We forget too readily that not everybody or anybody was free to marry, but that to establish a home was itself a privilege.

"PERSON"

We wage-earning masses are all too often without any responsibilities in the community. The marriage of two wage-earning youngsters does not alter much. How can anyone who is left irresponsible call himself a person? Officially, we still give him this title. But it is a purely honorary title. At the conveyor belt, in commuting, in punching the time clock, man is not a person, for he is uprooted and insecure; in his leisure his alternatives are too multiple to be called responsible. Wherever a community celebrates a real holiday, the members of that community act as responsible agents. But when a night-shift worker spends the afternoon in a movie or a pub, at the racetrack or at the zoo or in his garden or guessing a crossword puzzle, when we dial the radio for one of a dozen programs, we do as we please. The choices are so numerous, so indifferent, that it would be the abuse of a glorious term to call these choices personal.

The dignitary who was called a person by canon law had received this tremendous name in the image of the triune God one in three Persons! The connection between God's Persons and our faith in being persons should forbid us to call ourselves persons by nature. If the social order does not reflect the personal life of God, it is useless to bandy around the concept of a person as though it existed in some realm of nature. "Person" participates in the bond between God and Man. In ourselves, we find everything but "personal" features. Stripped to the bone, postmodern man finds atavistic fears, childish dreams, senile deficiencies, animal instincts: to be a person, then, is nothing natural, but it is the process by which we have been so loved that we remain connected with God's powers of impersonation.

Now, from the Reformation to the two World Wars, the general trend was to expand the status of "person" from dignitaries to an ever vaster number of people. Renaissance artists and scientists claimed "personality" in rivalry to the clergy and the princes. "Everybody is by nature a person," was the battle cry of the world for 400 years.

Some of us who live under the conditions of modern mass production may begin to wonder how this mere extension of the benefits of personality was ever held plausible. The majority, however, still lives under the spell of this dogma: we are persons by nature! Thus a temporary trend of extending privileges was exalted to "naturalness."

"NATURE"

This led to a second fallacy. For, the term "nature" now included the presence of the highest spirit in us. If we were persons by nature, "natura" became something infinitely bigger and better than it had been in the times of the living Christian faith. Modern man wanted to base his political claims on the tenet that "nature" contained the "person."

In pagan times, people had written on the nature of the Gods, *de natura deorum*. But the early Christians wanted none of it. God's mysteries were not to be treated in a "natural" discussion. In pagan

Corinth, people had mistaken the natural psyche for the personal life of a living soul in the Spirit. But St. Paul's letter rebuked these psychologists.

The Renaissance immersed man again in nature. Today, at its end, man is academically equated with his psyche. And God may be said to have a nature.

Against this, the Counter Reformation mobilized all its intellectual ammunition. The "supernatural" was apologetically defended. But our enemies mould us nearly always in their own image. In fighting them we ourselves become like them. Fighting a police state, we might establish one ourselves. Something like this happened to the Counter Reformation. The supernatural was defended with a certain success. But the "natural," in the textbooks of theology, became a copy of Renaissance "nature."

Now, this Renaissance "nature" not only extended its claims over "persons"; it also changed its quality from anything which *physis, natura*, had meant in antiquity. *Physis* meant "plantation" in Greek; Plato called God a planter or *physis*. The word comes from a verb which means "living growth". Physics, however, in the Renaissance, became what it is today: the science of dead matter. For the first time in the history of thought, dead matter was held to have preceded living growth. In a living universe, too, we may have to cope with corpses. But the mechanical "natural science" after 1500 tried to explain life out of its corpses by making nature primarily a concept of dead mass in space!

Only recently have we discovered that the term "nature" between 1500 and 1900 was used in a sense or with an accent unheard in any other epoch: mass, quantity, space, i.e., dead things, filled the foreground of scientific thought. Physics was held to "explain" chemistry, chemistry biology, biology psychology, psychology theology! Dead things were to explain the living. This new horrid degradation of the term "nature" itself made all personality values appear as the result of some drop of adrenalin in some glands.

Together the expansion of the term "nature" over "person" and "community" and the change in quality from living nature to dead nature made all apologetics of the Counter Reformation sterile. For they had surrendered to the enemy insofar as they shared with him the fallacious two *Metabasis eis allo genos*, the two denaturalizations of the fundamental term "nature."[1] A "human nature" once looked upon as primarily mechanical could not be restored to its splendor by any halo of the supernatural. The dichotomy was becoming too fallacious in its first half, "nature." If there was mechanically a "human nature" and if it was explicable within the nature of "physics" like any quantitative mass, it could then be handled by an ethics *more geometrico*.

This Spinoza-istic ideal of a mathematics for human conduct influenced the casuistry of the Counter Reformation for two reasons: the man himself could be thought out in advance, and the world could too. This means that when geometry is the proper approach to knowledge, God's creature "time" is murdered.

"TIME"

Good Catholics today think nothing of repeating the formula that time is a fourth dimension of space. This is a perversion by the Renaissance, as unwarranted as the perversion of "physis" meaning "growing life" into the "physical world" as meaning dead masses in space. With the pious pagans of antiquity, all time was rhythmical. When the Church in her hymn praises God *"qui temporum das tempora"* (Sunday Lauds), or when the liturgy prays *"et in saecula saeculorum,"* this rhythmical pre-Christian experience of living time is being shared. Before 1500, time is rhythm and cycle, musical interval and seasonal recurrence. Never before the Renaissance was time conceived as rectilinear, as "natural," or mechanical or geometrical. All the temples of paganism expressed the living quality of time by their architecture. Solomon's temple was no exception. The 365 days of the year were depicted by its measurements. Time was harmonious movement, not a quantitative accretion.

But with the Renaissance, this changed. Time was degraded to a concomitant of dead masses in space, it was no creatura, had neither rhyme nor reason. It became a mere quantity. Descartes himself was frightened by this result of his own principles. So he said: God seems to create time in every instant anew!

Again the Counter Reformation fought this enemy by insisting on the "eternal." "Eternity" was set up against this dead "natural" time.

---

1. *Ed.* The Latin phrase refers to an unacknowledged change in the basis of an argument.

But it was with "eternity" as it was with the "supernatural" against a wrong conception of "nature." Time once falsely conceived is not cured by eternity. All our traditions of time, pagan, biblical, ecclesiastical, had contrasted the eternal with the eons of eons, the *saecula saeculorum*, the succession of human generations, the *temporum tempora*. In other words, we lose our access to the eternal if we contrast it to that fallacy of classical physics, a non-rhythmical, dead time.

Several misunderstandings arose. The calendar of the Church, e.g., depicted in its one year the thirty years of Jesus' life and the millennia of the Church. Fifty-two thousand Sundays before God are as one day. Therefore God's six days of Creation in Genesis had never been analyzed as to their length. Now, the six days were taken literally. Man's own long life of seventy years lost biographical significance. The fear of the wrong time of physics drove the theologians to desperate attempts either to mechanize "eternity" as though it were a mere idea, or to persuade the faithful that the one-year calendar of the Church really gave in itself sufficient room to the miraculous birth of their souls. Alas, our souls do not unfold in one year.

Miracles themselves were apt to be thought of as exceptional invasions into the mechanical entropy of space by the Eternal. But God is no exception: He is the ruler. We are His miracles either always or never.

"MODESTY"

The degradation of miracles into exceptions was not yet the worst. The worst consequence of our "killing time" concerns our sense of *modesty*. During the last centuries we nearly forgot what shame was given us for. Shame is the soul's garment against arbitrary and untimely knowledge, because timing is the condition in which alone the Eternal may be revealed. It takes time for a bride to know her love. It takes time for a nation to find her destiny. It takes time for the heart to know itself. The mind of modern man whispered instead: it takes no time to know anything.

The Counter Reformation tried to save our chastity. It was felt that our secrets must not be unveiled too early. But the men of the Counter Reformation shared too often the prejudices of rationalism that the mind's knowledge was timeless; hence they changed the quality of our power to blush: The Counter Reformation cannot escape the reproach of having become prudish.

For instance. It was said by the promoters of Aloysius Gonzaga's canonization that he never looked at his own mother for fear that he might see the woman in her. In such a statement the secure faith between children and parents is destroyed. It is replaced by an onion-like scarecrow of a female body dressed up as your mother but remaining a female to you just the same. But my mother is my mother. How can she be anything else? The method with which the young prince is credited, which was not to look at his mother at all, makes the healing impossible. This then is the prudishness of the Counter Reformation. I do not judge the young Gonzaga, but I do criticize the proceedings of his canonization. Prudishness enhances the fear of obscenity instead of making it totally disappear. Prudishness never regains Paradise but makes it lost forever.

In Gonzaga's case, the recommendation runs that by not looking at his empress' and his mother's face, he rejected the flesh and became an angel on this earth. But the angels play before God's countenance. And the countenance of my mother is my first yardstick for chastity and shame. The faces of those who love us are as inextricably tied to our sense of shame as eternity is to true rhythmical time or as the supernatural to the living garden of "natura" in the Greek sense.

The Counter Reformation, by separating our mutual beholding, our countenances from our modesty, destroyed the "biological" time for shame. If shame is not the expression of growth, it turns into a loveless, asocial, hard and fast thing. Shame is our rootedness inside the garden bed formed by human countenances; it circumscribes our real life. And though it may come to some readers as a shock, our real life requires the experience of loving faces fastening on us. God's countenance cannot fasten on us unless His delegates, loving faces, are recognized as gateways to His face. Through them, we become unashamed members of that family which depicts the living God. It is easy to say that we are made in His image. When it comes to believing this stupendous truth, we must proceed in mutual convergence, or all our sayings are eyewash.

We may pause and take stock.

We have taken the term "nature" out of its context in physics because dead things are corpses; but nature is alive.

We had to take "time" out of its context in physics. For, in physics it is a dead preconceived quantity. Time is alive, rhythmical, cyclical.

And now we are taking "shame" out of its context in human zoology (called "psychology"). For in human zoology, shame is a guilt complex around sex. Shame is, in truth, the mortar of our edification into one living temple. The living stones of this temple must look at each other, must face, comfort, countenance, illuminate, view, regard, respect, perceive each other in perfect freedom.

We must therefore liberate the words "nature," "time," "shame," from their dungeon in "physics," and we can do this only by polarizing their light again. Time as opposed to eternity is the living eon, the cycle of our times. Nature as opposed to supernatural is living, sprouting growth. Shame, modesty, as opposed to self-revelation, is the custodian of the threshold of the time when we are to lift our countenance to a wider or deeper view.

The apologetics of the Counter Reformation defended eternity, supernaturalness, revelation; but it is astounding how far they conceded to their humanistic opponents the definitions of nature, time and shame.

We must repudiate these definitions and thereby emerge from the fall of the last centuries. This strange relapse into pre-Christian modes of thought occurred (I trust) for this very purpose, that this time we all, believers and unbelievers, might emerge from this fall, together.

Certainly no Christian can mentally remain in the abyss opened between physics and apologetics. Ideally such a man clings to revealed truth; materially, he rejects the experiences corollary to revelation: rhythmical, organic time, the creativity of shame as a gradual dropping of one veil after another, a living universe.

Was ever any Prometheus more cruelly tortured than these Christians were tortured mentally during the last 400 years by the spirit of the Counter Reformation?

Let us leave this Caucasian rock. Since every mind gave in to the death by "physics," the liturgy's healing power may also heal every mind, if only we will let its virtue flood our minds outside the sanctuary, too.

INDIVIDUUM AND EXPERIENCE

Lest the reader shrug off this breaking apart of our ideal and our material world, the treatment of two more terms as used in our day will amplify the contention that nothing but liturgical thinking can regenerate our basic concepts. For by the decay of these two terms, the present geopolitical crisis has occurred. The terms "nature," "time," "shame," may still be rated as purely intellectual terms. Not that they are; but few people are alive to the fact that their usage is of public significance. However, the terms "individual" and "experiment" dominate secular American thinking. A mistake in their articulation alters the public life of the people. Exactly this has happened! From good liturgical terms, individual and experiment have descended to the rubbish heap of the world of physics. Who even is aware that experiment and *individuum* are Christian terms?

"EXPERIMENT"

The "holy experiment" of the Puritans was not "an experiment in living." But life was a holy experiment. Hence the term is still bandied around. Modern man uses experimentation in education, marriage, friendship—i.e., everywhere where it does not belong. World Wars I and II were nearly lost because the people in the U.S. insisted on treating the event as a mere experiment.

When we hear of experiments, the modern mind thinks: this is a free choice, a situation of take it or leave it. If life were in this sense experimental, the necessary things, e.g., the Cross, Revelation, could have no place in it. As most people today are imbued by this spirit of pseudo-experimentation, religion becomes either an opiate or a luxury. For if we play around with everything, it sounds impossible that man's salvation should consist in his daily discovery of the "one thing necessary".

Here again, apologetics may "recommend" religion as a good thing, but it must remain sterile in an environment which does not see that our inspirations as well as our sensations are sanctified when they cease to be experimental: We are God's Holy Experiment. For we are in His crucible; God is creating us. The term "experiment" as used in physics is a poor second, a mere loan made to the laboratory by the language of the Church. The scientific term is a loan from God's proceedings with His children; and in these, the experiment is not arranged according to the theory of the physicist, but it is offered us by the love of our Maker who proposes to us and tests our degree of loving response.

In the laboratory, any experiment means the isolation of some elements for the special testing of a mental theory. God, since He is no theoretician, does not isolate us in His experiment. Quite the contrary. Whenever His experiment succeeds, a human soul gives up her isolation. When God experiments, He exposes us to danger (*ex-periculum*) lest our heart never wake up. When we experiment, we imitate His serious, unique acts of creation by our playful acts of research. Certainly, we too expose the materials in the crucible of our tests to danger. But we do it in mental pride: the guinea pig may die. God does not want the death of the sinner.

"INDIVIDUAL"

"Individualism" is the second term stultified by modern secularism. Even good Christians can be heard to start with "the *individuum*" as a given fact. The *individuum*, however, in St. Thomas Aquinas, has not at all a purely factual meaning. *Individuum* is a good Christian term as long as it means two qualities in one:

1. That which cannot be divided into smaller fractions by us: the atom.

2. That which we *may not* subdivide, i.e., the Trinity. The Trinity is indivisible. Peace is indivisible. "*In nomine individuae Trinitatis,*" peace treaties were concluded from 800 to 1815. The Treaty of Paris between the new U.S. and the British Crown began "*In nomine individuae Trinitatis.*" *Individuum*, since 1100, is a blend of the Greek "indivisible" and the Christian "indivisibility." And thus, enemies could make peace because their uppermost unity could not be torn to pieces by any war. God was indivisible. Today, when the indivisible Trinity is lost to view, Russia, Germany, the United States cannot conclude a peace. For they feel basically divided. How can they unite if nobody is being recognized as bigger than themselves? If no eternal God is capable of unity and at the same time demands our support of His indivisibility?

The fission of the atom, the schizophrenia of individuums, the twilight of the peace, go to show that anything can always be subdivided. The lack of peace, on the other hand, the horror of atomic bombs, the state of our mental health, prove that not everything *may* be subdivided.

"Individuum" has veered away from its polar meaning. Man, between 1500 and 1900, could be called an *individuum* because he participated both in God's qualities and in the world's qualities as well. In the middle between the atom and the Trinity, he boasted of "individuality." This *individuum* of the Renaissance boasted loudly in the face of the whole world: "I am unbreakable: I am impregnable." And Renaissance Man intimidated the powers that be so that they honored his divine triune likeness to the *Individua Trinitas*. Genius has been given his berth, through patents, copyrights and many other individualistic laws.

But when today Catholics and Protestants and Jews connive in the latest modern usage of *individuum*, they degrade the term *individuum* from a polaric and so to speak three-dimensional meaning. The latest usage of *individuum* omits the condition: that which *should not* be divided. It is impossible ever to form a community out of certified atoms. The sociology or history or economy of our Christian scholars very often is indistinguishable from the doctrines of the modern mind. The wrong acceptance of "individuum" as a given data makes it useless to patch up such chaos afterwards by recommending "social" measures, social welfare and charities. Man never, never, could live, breathe, speak, write or think, unless he is the image of the *Individua Trinitas*.

Thus, we have seen that in the terms "nature, shame, person, experiment, time, *individuum*," the apologetics of the Counter Reformation has made too many concessions. The antitheses of the supernatural, the social, the decent, the holy, the eternal, must all be misunderstood, once the theses against which they stood have changed their meaning.

To the artillery duel of such depraved theses and impotent antitheses our mind is exposed year in and year out. My own mind at least was. The theses I always knew to be wrong. The antitheses I always knew to be useless. This of course was purely negative. How my mind learned to become one, whole, indivisible again in its manners of thought and speech is a different story. And this story is the story of how the liturgy can become clear as the thread of Ariadne which leads out of the modern labyrinth, and makes the human mind again the temple of the living God.

II.

"O CREATURE MAN"

When a child hears its name, it is irresistibly forced to move. I can't hear my name without being moved in one way or another. Any potent love gives the beloved a new name, and by virtue of this name, he begins to move. Children, the overflow of their parents' love, move in their first appointed grooves because the name by which they are called creates their conduct, their movements, their walk through life.

When we grow up, the source of this creative procession through the first twelve or fifteen years of our life diminishes in power. We then must be loved again lest we cease to move in the right direction. Hence, a new name will enter our ear and fall upon our heart, the name by which God calls us to love Him with our whole heart, our whole mind and with all we have.

But we could not enter upon this new love had not old love foreshadowed it. This time, the parents who acted in God's stead when we were infants are not calling. The new call comes from outside this world. It reminds us that we enter a new home, the wide home of God's creation, into which He now sends us to represent His holy temple. All ancient temples depicted the heavens. But man depicts the Creator of the heavens. Among the other creatures around us we are to be created into His image.

It is at this point, when we are starved for our new name, fearful of the immensity of our new home, that the reprocessing of the creature man usually is retarded or interrupted.

For it is all very well to say that the love of our parents now has to be regenerated, that someone must love us now with a fresh and unheard-of name. In real life, there intervenes a long time of doubt and misery, of affamished despondency, between our breaking away from our mother's apron strings to the certain and elegant movements of our definite procession and progress. After puberty, the mind begins to tear the old home of ours to pieces and our body timidly tries to fall in love. After puberty, man prepares for the new love in two compensatory movements. He must make room for the new love; this is the mental process called doubt. He must admit that he has to find the bridegroom of his soul. This bodily process is his *Wanderjahre*. By doubt, we assert our independence from the old; by restlessness, our dependence on the new.

MENTAL INDEPENDENCE, BODILY DEPENDENCE

This is a protracted period during which the old home is not yet completely relinquished nor has the new one entered visible reality. Two movements overlap: the mental one tears down the visible old hull; the bodily prepares for the yet invisible new one.

This overlapping of the mind's tearing to shreds by doubt and the body's getting ready for a new foundation usually is misinterpreted as the mental and bodily processes of the natural man. By looking at them as a mere juxtaposition, logic makes its gravest mistake. The negative task of the mind and the positive task of the body are not seen in their *mutuality*. The growth of mental independence and that of bodily dependence are not seen as two aspects of the same growth. Mind and body appear as two fiery horses, instead of as a process to regain the next equilibrium through a wonderful balance between the opposite direction of mind (critical against his past) and body (desirous of his future).

The strange fall of Western man consists in the illusion that mind and body are not two compensatory and strictly time-directed processes of Me, but that they have being and that they tend in the same direction! The ridiculous parallelism of the physical and psychic processes was one of the many faith-less theories of the Renaissance.

A TRANSIENT PHASE

Any one of the theories on mind and body arbitrarily assumes that the student between, let us say, 14 and 21 is the model case of a so-called natural man. A *transient* phase was singled out as the *norm* of man's nature. However, the two outstanding proofs of its transient character are, 1) that the mind is our critical faculty to shred the past and to make room, 2) that the body in the meantime storms forward to our final destiny. The doubling of the mind by doubt and the halving of the body by sex are two sides of the same process. And this process is to make a child, a listener to old love, gradually into a speaker of new love.

Instead, Reason abstracted from this functional role of "doubt times sex" and postulated a free mind and a free body. This only goes to show the embarrassment of man. He tried to speak in public only of his mind's grandeur, and suppressed the miseries of the body. But what is wrong when we split inside ourselves into the listener and critic, the previous child and the future man—lest we remain attached to old shells of life—and at the same time shrink to one half of our bodily self so as to become worthy of our new attachment in life?

The man who hears this new and next attachment called out over him may speak with Romeo: It is my soul who calls upon my name. And at the same time, this same Romeo will excel in every form of sophisticated soliloquy inside himself, and by becoming two people inside his inner debate, he will break all the ties with a dead past. The experience of our mental double and of our bodily "better half" is one and the same experience through which our anchorage in past and future remains guaranteed, although, on the surface, we appear in this period as thinking individuals.

But before we can think, we owe thanks. Before we ourselves may reason, we have reasons to believe. When thanking and thinking, reasons and Reason, cease to be recognized as modalities of our life, we get stuck.

Natural reason is a very special reason sprouting in the unfulfilled mentality between 14 and 25. It is the Reason of the classroom student. Greek philosophy, eighteenth century enlightenment, American common sense or pragmatism, are gigantic superstructures of these uprooted minds and unloved bodies in their in-between age, when one set of names has faded and the new call of love is slow to resound.

The tricks of all these doubting minds and fallen bodies is to call their transient state the natural one. The nature of man, they claim, is vested in their mental and bodily processes. "Psychology is the science of the mental processes," the most famous textbook of American psychology begins. A textbook on physiology might echo this desperate division in mind and body by saying that physiology is the science of the bodily processes. This division poses as "scientific." Most good Christians and nearly all theologians repeat this residuum of pre-liturgical thinking daily by quoting approvingly the pagan adage "*Mens sana in corpore sano*"—the mind in the Platonic prison of the body!

In all these analyses of the "natural" man, the soul is at best allowed in as an afterthought. But the soul is incarnate. There is no body and there is no mind *per se*. The soul alternatingly uses either mental or bodily expressions to become incarnate in her earthly role. From soul to role via mental and bodily expressions, we shall "take place" and take our place.

If and whenever the loved soul is not directing mind and body, these two divided horses pull apart and simply "take place." But we shall not simply take place as a detonation. We are to take *our* place This distinction psychology and physiology, sociology and medicine, all have ignored.

The only place in which this process has never ceased to flow is the liturgy. From the liturgy, I have learned to think rightly.

"THOU" MAN

The first step was the discovery that we are and remain nobodies—*massa perditionis*—unless we are called by our names. In 1916 I wrote "A Doctrine of Higher Grammar" (printed in 1923) on the fact that we must move through every experience as figures of grammar lest the experience never be made.[2] The soul must be called "Thou" before she can ever reply "I," before she can ever speak of "us" and finally analyze "it". Through the four figures, Thou, I, We, It, the Word walks through us. The Word must call our name first. We must have listened and obeyed before we can think or command. This, then, is the health principle of the soul. When addressed by the Spirit we are the liturgical "Thou." This takes precedence over our other liturgical shapes such as the Ego, Us or It.

In the liturgy, this sequence is revealed. The first figure in our liturgical treatment is "Thou." The priest then only is allowed to

---

2. *Ed.* See Rosenstock-Huessy, *Practical Knowledge of the Soul*, trans. Mark Huessy and Freya von Moltke (Eugene, OR, 2015).

respond "I" after he has been called out, in his ordination, by his full name and has made the sacrifice of his will. The congregation led by him is able to receive its historical religious role as "We." And at the end of the service, the objective statement may come: "In the beginning was the Word."

"THOU" CREATURE

"Thou," our first liturgical shape, although called the second person in grammar, is by no means restricted to us humans alone. Creatures which never may say "I" or "We," may yet reach the first phase of liturgical life, the phase of "Thou." But we, too, must all our life long stay in this phase. Even the whole Church must remain God's creature. *Creatura hominis* means "O Thou yet to be created child of man." Strange as it may sound, there is no salvation unless we return into creation. Thinking man is only redeemed by thanking God again as a creature. The recently deceased Joseph Wittig taught me this. Two decades back, he edited a quarterly, *Die Kreatur* (The Creature). For its second volume, he translated the rites for blessing salt and water:

> I adjure thee, thou creature of salt, by God who commanded thee to be thrown into the water by Eliseus the prophet in order that the sterility of the water be healed; that thou become consecrated salt for the salvation of the faithful; that thou be for all who drink thee health of soul and body; that thou put to flight and drive from every place in which thou art sprinkled all fallacy and wickedness and cunning of seductive illusion and all impure spirit; thou art adjured by Him who will come to judge the living and the dead, and the world with fire. Amen.

Wittig added:

> In the same manner, the Church also speaks to the creature of water, and it certainly is no accident that she does not speak to salt and water but to the *creature* of salt and the *creature* of water. Salt and water cannot hear what we say. They can only react chemically. Once taken out of Creation, they are dead and deaf and don't react to Word and Spirit. And they are immediately taken out of Creation when they no longer

are spoken to as creature. When they are spoken to, they stand in the realm of the "Thou," where there is life and listening; otherwise they are in the realm of the "It," into which neither speaking nor harkening can reach.

When they are spoken to as creatures, they are spoken to in the faith, therefore in "virtu," and are sought and met in the living hand of God in which nothing can be dead or finished, but only living and becoming, where, for this reason, everything still is miraculous. "When the Church says: 'Thou creature of salt,' perhaps she says so in order that she may address the salt at a moment in which it still is in the miracle-working hand of God as much as is the Church herself."

Cor ad cor loquitur, Cardinal Newman's motto runs.[3] Under which conditions heart speaks to heart, Wittig tries to ascertain. Not when we worship a teacher or adore a woman or are spellbound by a spellbinder or enamored with art or are crushed by venerable authority does the heart speak to the heart. When Creature speaks to Creature, then *cor ad cor loquitur*. Only then have we been cleansed of our mental idols. For God has given us a heart of flesh; and no raging against the flesh may prevail against this fact. Few people understand this as the law of our intellectual life; the liturgy lives this law.

PROCESS OF CREATION

Some years ago, I attended the blessing of the wine on St. John's day, at St. Paul's Priory in Keyport, N.J., with Dom Thomas Michels officiating. In those unforgettable days of Christmas, I was struck with the formula *"Benedico te, o creatura potus—*I bless thee, thou creature drink." This pushed my understanding beyond the rather simple act of substituting the word "creatura" for "natura." It dawned on me that there were before me not two interchangeable words but two completely separate tongues. "Natures of things" abstract[s?] from the historical hour of revealing speech. "Creatura potus" demands a definite renouncement of abstractions. Why? This drink has passed through many stages; some of them common sense assigns to "nature," like planting the vine, pruning, fertilizing, spraying with sulphur, etc., etc.

---

3. *Ed.* John Henry Newman (1801–1890), English theologian and major literary figure, who moved from Anglicanism to Roman Catholicism. He was a favorite of Rosenstock-Huessy's.

Some others commonsense classifies with social action, like the harvesting, barreling, bottling, etc.

But in the light of "creatura potus," our academic distinction between natural and social collapses. In the sight of God, *we* His faithful—when we have done right by the wine—lead this creature wine as much to its destination as does the soil, the rain, the air, the sun. We are not God, but one of the creatures in that meadow of God on which all creatures here below praise Him, since they can't go about their business without fostering at the same time everybody else's business.

In other words, whenever mortal man leads the other creatures to their destination, we do not prevent, rather, we complete their "proceedings" into that Creature which is in process of being created.

History, social processes, human mores of the vintner or the salt miner, or the pump builder's techniques are not to be thought of as arbitrary, willful acts. They may be steps in the continuation of creation itself. To call any element "natural" is not a description but an act of decision. What you call natural, you exile to its own beginnings. No wonder that after 400 years of natural science, the analysts trace us to our mother's womb. This last outcome of the Renaissance is not fantastic. Only that this is called "science" is fantastic. When I take a sledgehammer to crush a tin can, I am not being scientific. But the term "natural" is precisely a mental sledgehammer which reduces this glass of water, "O creatura aquae," to the Hades of H^2O.

"CREATURE" AND "NATURE"

Now the Counter Reformation, though not altering the liturgy, has confined it to its narrowest place at Mass. The greatest recent commentary on the liturgy of the Mass has not even room for "creatura" in the index. Instead, the objects of the Benedictions are called "naturalien." This German term is hard to translate. But to be called "Naturalien" gives the blessed gifts the poorest possible rating. They seem to be taken out of the palms of their Maker, they are dead on the ground.

I do not think that the incarceration of wine, salt, bread, fire into mere "nature" makes sense. Why should we ever call them, in theological or philosophical or historical books, natural things? I have given up thinking so unliturgically of the universe around me. If the so-called scientists must do this, well, in Armour's stockyards in Chicago the cattle are killed by a blow against the skull. But for the fire in the living oxen (Ezechiel 1:13) the stockyards don't care. To naturalize is a second-rate function and a mere afterthought.

In a narrower area, the term "naturalia" for the creatures which we may "Thee" and "Thou," blinds the liturgist to the fact that the catechumen, the "echoing" beginner, is on the same level as the creatures who only may hear but not speak. Each Mass recognizes that there are four degrees of "speech": There is one who only speaks. One who speaks and listens. One who listens and responds. And one who only listens. These degrees of the Word are held by God, Clergy, people, catechumens; they occupy these four levels of speech. And in every Mass, the word who was in the beginning unfolds again His four-terraced cataract.

THOU CREATURE MAN

If wine, if a drink, harvested, fermented, bottled, may be "creatura," then we may trust that man, despite the protracted phases of his social climb in the midst of his social groupings, has not necessarily lost his power of being a creature in the process of creation. Such a man walks according to God's word unto his own final "creatura," as do salt, water, fire, wine. He need not get lost in the concepts of *natura et supra naturam*, times or eternity, person and community, shame and countenance. But in this walk he is no longer the *homo sapiens* of zoology or "psychology". For "homo" may signify the man who has no throne of love, no appointed groove, but is dethroned and disappointed.

Through the genitive in "creatura hominis" a direction is given to the dead term *homo*. It is mysterious enough. *Creatura* is a form of the verb which points to our final appointment, to the vital form "venturi Saeculi." The phrase *creatura hominis* puts man, so to speak, into his teleological genitive. As *creatura hominis* I am addressed as that part of me which is yet to come! *Creatura hominis* is an act of faith, as it places you on the side of destiny. *Natura hominis* is an act of destruction. It changes man's status and does not even admit that it does anything. The refinement of the devil seems to be just this, that he pretends to do "nothing." In fact, he annuls. If this wine, now blessed, after so much honest toil and sweat of its growers, is still "mere" nature, then part of creation is annulled. Natural science annuls! The time of God spent on His beloved creatures is annulled.

Time is regained for your future whenever *creatura hominis* is called out over you. You are now not "natura," because you are not who you are but you are told who you are going to be. You are not determined by the clocks of earthly and commonplace time, because a new era opens with this next step into creation. You are not a responsible person but the responsive plasma in the hand of your Maker, and your bashfulness, your shame is overcome by the veil-removing power of a new birth.

Creatura hominis—it will take another article, yes, and a whole long book, to show how the Mass expresses this transubstantiation from *natura hominis* to *creatura hominis* in its whole formation, action and sequence, how the steps taken in the Mass to lead the faithful through their grammatical modes unfold the only valid logic of reasoning.

Enough has perhaps been said to justify our claim that the liturgical rebirth signals the birth of liturgical thinking for the creature man, the *creatura hominis*, the child of man yet to be created.

$$\text{Eight}$$

TIME BETTERING DAYS

A Paper Read in the Tucker Fellowship at Dartmouth College on the Spring Equinox of 1954. It was published in German in *Die Sprache des Menschengeschlects*, I (1963).

TIME BETTERING DAYS .... The learned say that this phrase is found in the English language once, and once only: in Shakespeare's 82nd Sonnet. And the unlearned say that it is a contradiction in terms. How can time which is the totality and the compound of all days be improved by days which are better than the rest? This really is pulling time up by its own bootstraps. It can't be done. Time bettering days are nonsense.

Having published my first study on the calendar in 1910 and my last in 1952, I am taking this highly unexpected summons to replace our friend Booth, the impressive reader of Shakespeare, as my opportunity to reorganize my calendaric and time studies around William Shakespeare.[1]

Shakespeare's and my own thesis is that mankind in the Christian era lives in a pluralism of intersecting and overlapping calendars; that this is the distinction of the Christian era as against all others, Jewish, Chinese, Roman, Aztec and all the rest. Our era is the era of pluralism in timing, and this is officially its character. Jesus said to the man whom he saw working on a Sabbath: "Man, if you don't know

---

1. *Ed.* Edmund Booth (1895–1974), professor in the English Dept. at Dartmouth, lectured on Shakespeare.

what you are doing, you are cursed. However, if you do know why you are working you are blessed", and this word of the Lord of the Eons of Eons has become flesh in the pluralism of every Western Man's own eons. His times ever since have been allowed to become a spectrum just as waves or colors. And only all the colors of the spectrum experienced together create the fullness of time bestowed upon us in this eon. We all live by these new standards. But strangely enough, few seem to know this fact. Not the whole spectrum of the times can be occupied by us this evening. I have given it completely in a whole book.[2] Not even Shakespeare's whole spectrum of time may I spread before you. I tried and the paper was three hours reading time.

Even as it stands now, this paper probably may have to be too long for your liking. And thus, you may tend to label this evening a time worsting, a time deteriorating day. Even in this case, as we learn best by contrast, you would still benefit and increase your understanding of time bettering days. By boring people, we may arouse their own intensity, and each birth of Time = Bettering Days is dependent on an enhanced intensity.

There are three parts to this paper and an introduction. The introduction shall remind us of two treatments or aspects of time in the work of Shakespeare. In their light, part one shall treat the academic calendar and our own College Times, the second part deals with the origin and the change in meaning of one English word—the word "Noon", and with the calendar implied in its strange triumph. The third part shall deal with our free, unpredictable, and biographical High Times suggested by Shakespeare in his Merchant of Venice. The platonic, the medieval, and the biographical calendars are going to claim a share of your allegiance to timely living.

The Introduction to our Theme naturally is:

Sonnet 82

I grant thou wert not married to my Muse,
And therefore mayst without attaint o'erlook
The dedicated words which writers use
Of their fair subject, blessing every book.

2. *Ed.* Rosenstock-Huessy wrote about "time" over many decades. Much of what he had to say was collected in vol. 2 of his *Soziologie*, "Die Vollzahl der Zeiten" (Stuttgart: W. Kohlhammer, 1958. See also *Soziologie* in the Bibliography.).

Thou art as fair in knowledge as in hue,
Finding thy worth a limit past my praise;
And therefore art enforc'd to seek anew
Some fresher stamp of the time-bettering days.

The New Oxford Dictionary of 1912 cites this single example
of the compound "time-bettering"; only once more has Shakespeare
himself come near to this expression, in Sonnet 32.

If thou survive my well-contented day,
When that churl Death my bones with dust shall cover,
And shalt thy fortune once more re-survey
These poor rude lines of their deceased lover,
Compare them with the *bettering of the time* . . .

In this phrase of Sonnet 32, however, the power which does better the time, is not mentioned. Only in 82 is the fraction of time, the
day, made to install progress into time; this is the most radical faith in
progress in any linguistic expression known to me.

However, Shakespeare opposes this faith in time-bettering times
by a very opposite hope: eternally recurrent time. You are all familiar
with Henry VI, the boy King's, sigh:

O God! methinks it were a happy life,
To be no better than a homely swain;
To sit upon a hill, as I do now,
To carve out dials quaintly, point by point,
Thereby to see the minutes how they run,—
How many makes the hour full complete;
How many hours brings about the day;
How many days will finish up the year;
How many years a mortal man may live.
When this is known, then to divide the times:—
So many hours must I tend my flock;
So many hours must I take my rest;
So many hours must I contemplate;
So many hours must I sport myself;
So many days my ewes have been with young;
So many weeks ere the poor fools will ean;
So many months ere I shall shear the fleece:
So minutes, hours, days, months, and years,
Past over to the end they were created,

Would bring white hairs unto a quiet grave.
Ah, what life were this! how sweet how lovely![3]

Compared to the Sonnets, this time is not progressive. It is not gloriously living forward, but it has one great merit: It would be circular and permanent and it would be universal. Most modern businessmen recognize in the king's yearning their own commonplace idea of what time should be like. And more serious than that: In the year 1918 at the end of The War, workers paraded posters in Paris, London, Rome, Amsterdam, Berlin carrying these strange demands:

8 hours of work

8 hours of sleep

8 hours of leisure

This was their idea of the World Revolution! Here, the vision of Henry VI had become revolutionary action. And do not say that the workers spoke of one day, Henry of all life. It is some perpetual human necessity of which you will find examples during this whole paper that the single day is used to express the whole principle of timing. Labor's calendar is indeed hoping for a permanent employment, an eternal chain of days of 24 hours length. And the simple day is this eternity's condensed experience, and therefore the 24-hour day is apt to serve as a slogan. This vision, then, has become flesh in our business calendar. Our calendar of production rules our factory system. And that is, it is accepted by both, capital and labor, as their manifest destiny. Variations like a 40 hour week do not alter the underlying principle. It is the dream of the Chambers of Commerce and of the Unions: make every Sunday, every holiday, make every Easter predictable through the next thousand years. Make capital pay interest in 13 months, all of identical length, for centuries; make the working week 40 hours, 48 hours, 32 hours long. Provide playgrounds for the leisure, provide dwelling boxes called better homes, for the sleepers. Make the number of jobs and of the unemployed army of industrial reserve constant. Don't allow them to fluctuate. Balance, by birth control, the times of youth and old age in overpopulated India, Japan, Italy. Count the 24 hours of the day in the abstract over large standardized regions called Eastern, Seaboard or Mountain Time. Swallow up all the concrete specific empirical observable noon hours when the sun

---

3. Henry's Monologue: *Henry The Sixth, Part III,* Act II, Scene 5.

is in his zenith, in favour of the general and theoretical whistle that blows and sends us all scurrying for our sandwiches. Disconnect the precalculated social time from its sensual origins in the sunsets and sunrises and real moonlights at the place of your actual habitation. Disconnect it also from the religious calendars with their "ridiculous" movement of Easter. Better the spaces in which men work, play, live; build bigger and better stadiums. But do not better the times. Predict them, instead, by Gallup polls. Treat space as your social medium, but time as an element of nature.

For us on the unreal campuses of education, this business calendar is not quite available. Even over those college professors who believe in dutifully reading their newspapers eight hours a day in their offices, the iron calendar of industry has only a fictitious or borrowed power. They may imitate the business man, but colleges live under an antithetical calendar. We really try to run this nation by factories and education, and that means, by two opposite calendars. This dualism is a fact. And this calendar of education I now shall investigate. Since Henry VI has his calendar incarnated in Labor's and modern business man's way of life, could it be that Sonnet 82 has foreshadowed our own college calendar?

At first thought, this seems improbable. A liberal arts college is an academic institution. Academic manners trace their origin to Plato. It would seem natural if our timetable owed something to Plato's calendar as he advocates it in the Laws, in the Epinomis, in the Republic. Let us look into this a little more. It is not very well known that Plato did ascribe Supreme importance to the calendar and demanded our total subjection to it. But such is the case. Plato sees man's toil as a laborer, as a warrior, and this is his earthly cave. Into this cave, however, the stars shine. Let me quote Epinomis 978: "Whether you call it in your pleasure Kosmos, or Olympus, or heaven, its revolutions constellate the stars, and they in turn give us the seasons of our nurture. How do we learn? If by numbers, how do we learn numbers? No other single feature is more beautiful than to watch the coming of day and the coming of night. What a reward when we continue doing this? Many nights, many days, the heavens send us so that finally even the man with the lowest I.Q. must understand. This is one reason for God's Creation (poiesis) of the moon as she in 15 days waxes and in 15 she wanes. And the 30 days teach us what a circle should be like." Or this: "An alleviating ointment to us poor mortals comes in the form of our joining the stars in their dance on our festivals and holidays."

Plato hoped that his Greeks might soon return to the Egyptian worship of the stars as their true guides, and that his perfect city would live in a permanent calendar of star festivals. This celestial calendar dominated even the statistics of Plato's city-planning. Neither economic nor military considerations interested the founder of the Best City. Instead, the twelve months of the year and the twelve hours of the day dictated to Plato the right size of the Citizenry. Every hour and every month some citizen had to observe the worship of the Gods in the sky. Now 12 x 12 equals 144. Plato concluded that his city must have 5040 citizens because 5040 can be divided into 144 by 35 observing crews. This starlore is truly absurd. It shows you how far away Plato's thinking strayed into Egyptian darkness, in a gruesome or desperate attempt to retard the coming of Christ. For, to Plato the divine logos is number, and number's capacity to count the many impersonal things of dead nature over awes him. However, to Israel, old and new, the Divine logos is quite a different power. It is the power to rise above numbers by bestowing unique names within the city of man, "Where move in strange democracy the million masks of God" (Chesterton) and address each other by name reciprocally.

The Platonic city with its 5040 citizens puts to death anybody who has private devotions, private teaching, personal lyrics, individual philosophy. Socrates was then the first whom Plato would have had to execute in his city! Plato went so far [as] to demand that his city should repeat this calendar without the slightest change through eternity. His Utopia reminds you of Hitler's proclamation that he would rule "Four thousand years of recurrence", or of the "China of the four thousand years in the Mandarin Rule of the Sky." If the leaders leave Plato's city on an errand to other cities, he writes, "let them afterwards tell their citizens that their own city is by far the best, better than any of the cities visited." Here Plato anticipates the modern Ministry of Propaganda. As in "1984" then, it shall be in Plato's city; we all dance and sing when the Commissars say so.

The greatness of this monstrosity lies in two facts: First, the cosmic order acts as our teacher. Its year is a revelation. And this means that Plato's City is purely educational; teaching the natural order is its highest goal. Second, the gods themselves will unite with the men and women who so learn. Celebrating the festive days means to feast and to cohabit with the gods. Again and again Plato assures us that the gods enter among the singers and dancers as their companions. Hence, we begin to understand—the queer number of citizens 5040 is not chosen for utility, but it is chosen for the apotheosis, the deification of educated man. Apollo and the Muses will come down from Olympus insofar as men observe the celestial rhythm. By forgetting the earth totally the Platonic City meets the gods halfway. Serious toil is one half of our fate; it represents earth. Dances and songs are the representatives of the Gods. Plato's error lies in the fact that earth and heaven remain eternally separated. But toil and orgy both are our human share. And this dualism is truer than Plato's dualism of stars and streets.

But what shall we say of Dartmouth in this respect? Has the modern academic world imitated Plato? Certainly, our calendars have only two short periods of serious toil. They are the examinations periods. Now, these come twice a year and they indeed do depress the soul of man, in a liberal arts college. But these two periods really are reduced to the bare minimum of earthliness. They so to speak connect the flying balloon which is a college, with the earth just enough so that we may appear in the eyes of the alumni to be a captive balloon and that we may, by this captivity, continue to get from the world the material support we require. The examinations assure the working community outside of our usefulness, seriousness, effectiveness and of our evaluability in dollars and cents. The examinations throw the necessary sand into the world's eyes. Otherwise, we really live in a heavenly sphere of papers, marks, assignments, awards, competitions which we have labelled, education. But it must be emphatically said that we have not adopted Plato's Calendar. Plato would have us kneel before the Equinox of Today, March 21. We don't kneel. We live in a second world of good thoughts, of Paideia, of education. The Christian Era, fortunately, has not allowed us to put faculty and students under the ironclad and relentless repetitions of the wheel of the firmament. What, then, is our calendar like? I give you its secret when I ask you to count the month of your academic year not as September or October, but as the first, second, third, fourth, fifth. sixth months by the enumeration of ordinals. Of course, you will object. You will point to the printed page where the time-honored names March, April, May do appear.

It is true such a calendar of Dartmouth College is given in the directory. It probably has given rise to speak of the mechanism of a college. I will give it the lie. This printed list does not tell the story of the real times of a Dartmouth man. It is one half of his true calendar, so much so that it only makes sense when you combine it with the unprinted—and often unprintable—other half of his same man's calendar. After all, the calendar is the timetable of our real habits.

When you, however, contemplate a timetable on which the two halves are unified, then you will convince yourself of two important points in the college's living processes: I. It successfully balances the business calendar. II. It supersedes the rigors and horrors of the Platonic tyranny.

Sept. 21	Registration
Nov. 23 Tuesday	Thanksgiving Recess begins
Nov. 29 Monday	Thanksgiving Recess ends
Dec. 18	Christmas Recess begins
Jan. 5	Christmas Recess ends
Jan. 24 – Feb. 2	First semester exams
Feb. 3 – Feb. 5	Mid-year recess

Now, the spices of these four months, not mentioned on the printed page of the Directory, are, of course, the football weekends and the house parties. The football weekends are subdivided in those for which you are expected to leave town, and those which take place here. The gods feast each other in eternal alternation. But we cannot feast Harvard or Yale here in Hanover. And there is the rub in the pulsation of the blood in the body politic called Dartmouth College, during its first two months. For this reason, it is misleading that we do speak of September, October, etc. To him who lives this calendar on the inside track the academic life begins on September 23 with a first month, to be followed by a second, a third, and a fourth. The four months of the first term, then, build up to a count of ordinal numbers. The numbers may be enumerated in a technical language of astronomy as September, the ninth month, and January, the first month. But for an understanding of the academic year, this puts a wrong construction on these months. Also, these first, second, third, fourth, months are periods which do not begin on the first of September or the first of January. Instead they begin quite independently from the astronomer's timetables when they begin: For example the first month of the college stretches from September 23 to October 23. The fourth month stretches from January 5th, the return from home after Christmas to the Midyear Exams at the beginning of February. At this point, we have the chaste expression, Midyear Recess. The true term is, Winter Carnival. It towers between the two halves of the year and makes them bearable. But more than that: It is Dartmouth's most famous contribution to college life at large. It has, for two decades particularized the name of Dartmouth more than the football matches. We do not know how long this carnival may preserve its unique character, with Colorado, Idaho, Middlebury competing. Suffice it to claim right now this calendaric significance for the Winter Carnival; by its celebration, the College acquires a vaster environment, a larger territory than during the rest of its annual course. The Winter Carnival relates the College to regions not even tapped at Commencement to quite the same extent or at least not by the same network of capillaries and contributaries.

Now, in all calendaric considerations, it usually goes unmentioned that times build spaces. The present day heresy of geopolitics, of regional planning, of space ships and stratospheric flights construes spaces independent of times. So-called Christian ministers preach neighborly love and thereby understand the people next door. They preach neighborhood dances while the good Samaritan treated the man of the next hour as his neighbor. The Winter Carnival is on the side of the neighbor of the hour, i.e., on the Christian side. For it creates a space of its own around Hanover; it creates a territory within which Hanover is located and therewith, the Carnival does what holidays should do: It makes times the masters of spaces. During a time span like Winter Carnival or like the Olympic games, the local differences and distances disappear. Contrast this with the Christmas recess and you will recognize the wide application of this law that holidays determine spaces. For, at Christmas, the map of Dartmouth as a nationwide college nearly is allowed to disappear. Why? Because everybody goes home. The individual student is sucked back into the place of his own nativity. The people who surround him at Christmas are his pre-academics, his pre-Dartmouth friends. At the winter solstice his past rises once more to swallow him up. The character of a holiday then is indeed very much contained in its group-forming quality, in its power over spaces. Winter Carnival is a strong holiday because it cuts through all the pre-Dartmouth lives at home. Also, it demands a very long preparation, and suddenly, our two American calendars stand revealed as space and time, earth and heaven in two opposite orders. The factory calendar places space over time. It believes in space bettering production. The educational calendar places time over spaces; it is itself a time-bettering day.

I have stressed Winter Carnival as it is not featured in our printed calendar. But the second peak, Commencement, is fully recognized by the authorities. On Commencement, the mountains parturiate and beget the rhythm of life. Commencement seems to me the most formative event of the American scene. It would be a very long story indeed if I were to dwell on its impact on our civilization in detail. May I make its importance clear by one aspect of which few people become aware but by which the behaviour of the millions is determined. This aspect says: What I do between June and September is of my own choosing. I can go to summer school, work on a construction job, travel on a freighter, bicycle through the 26 countries of Europe, make hay on a Montana million-acre ranch. But whatever I do during these four months, whether the Navy takes me to Labrador or my mother takes me to the West Coast, or I save money for the next year by hard work, these four months are my avocation, my individual's individualism so to speak, while the other eight months at Dartmouth are my individual's compromise of playing the game according to the rules.

To the majesty of the academic calendar, the Churchmen themselves have learned to bow. To give just one instance. The Clergy of any Episcopal dioceses meet with their bishop once a month at the so-called "Clericus". There are, however, only nine such meetings during the year. From June to September, the Clericus does not take place.

Since the Anglicans respect the liturgical calendar more than most other denominations, this their bow to the purely secular calendar is a telling symptom. No inhabitant of the United States can evade the influence of the partition of the year by Commencement and Labor Day.

When I discussed the work service with the Friends Service Committee, they decided to run their camps during the summer months. I decided to start mine on December 1.[4] Their's is an avocation; ours was serious. Their's was a net gain and a pleasure. Ours was a sacrifice, an uphill struggle and grim indeed. Our winter hardship, however, moulded every one of its members into new people stamped totally by this total plunge. The usual summer work camp's function is fulfilled when it leaves pleasant memories behind. Pleasant or incisive: That is the difference between the 3 and the 9 months. So much can the same external action depend on its calendaric context. And while the role of the context of words is known to every lawyer, any man, the context of life seems not to be known to religious and political leaders and to educators. Context is the key to the fabric of words; it also is king in the texture of our temporal order. An act done between June and September, and the same act done during the college year, differ as far as play and destiny, as security and danger differ.

We prepare the young for the serious life and at Commencement they are delivered to the world like doughnuts or *croxignolles* in due course. The Christian era creates bodies of time, because it recognizes every time by its fruits. The pregnancy of the alma mater from September to June begets the alumnus. Since many people misunderstand the revival of antiquity, it cannot be overemphasized that the most Platonic institution of the universe, the Liberal Arts College, has a Christian Calendar of Fruitfulness. The adoption of this calendar has been the tacit condition, the Christian condition under which the "ivory tower" of Plato has been re-admitted into our era. The revival of antiquity by the Renaissance took Plato into the climate of our own era. Our love, a fruitful love took into its orchard even Platonic love, that is fruitless love, and allowed it the place that even fruitless love deserves, in the preliminaries of life, in education. But woe to people that forget or deny this permanent condition of Plato's readmission. Plato's calendar and all it signifies—castes, slavery, lack of privacy— remains excluded and stands condemned. Again, we owe to Shakespeare the appropriate terminology for our modern treatment of a period of education. Hamlet speaks of showing the age and body of the time his form and pressure. Indeed, long before educators fill the

---

4. *Ed.* An allusion to Camp William James, which opened in the winter of 1940–41. See Jack Preiss, *Camp William James* (Norwich, VT., 1978).

classrooms with assignments, they already have created a structure, a form of time, by the lay-out of the academic year. The "pressure" of our final cause, of Commencement, may be said to transform the cyclical Platonic academies into pregnant bodies of time. Time, thanks to this pressure for fruitbearing, acquires a new, non-cyclical, progressive quality. This time shares the secret of all redeemed creatures: It is expected and it is promised and therefore it can be fulfilled. Of these three qualities living bodies of time are composed, and this was unknown to Plato. Time to him was rhythm and revolution and circling in endless repetition from beginning to end. And the academy was a world of its own endless repetitions, from 387 B. C. to 529 A. D.

Our schools have filled the Platonic contents into the Biblical streamed of promise and fulfillment, by grafting the promise upon the academy. We live from the end to the beginning because our destiny has been revealed to us. There is, I submit, in every single educated person's life a secular analogy to the Biblical "Promise" of taking us all together out of Egyptian darkness. But don't we see the glowing eyes of the young who are offered this opportunity? These eyes to me at least defy the refined distinction between the biblical and the secular. I would not know which word to apply to the great expectations of an eager boy; sacred they are, without benefit of clergy. It would seem, therefore, that our colleges have inherited a lot out of the complex inheritance of the ecclesiastical polity. We possess this one element of her revelation: a body of time lived not for a livelihood, not lived for a standard of living, not lived for good works or for work or for service; no, a body of time lived under the expectation of the novice, under the promise of the experienced, under the pressure of fruitfulness, a body of time pregnant with meaning because pregnant with promise, a beginning enlightened from its end.

And in this sense, the college calendar has a lubricating effect on our mechanized society and its factory system. That the college calendar has an allegedly secular character, does not alter the fact that it combats the mechanical character of the industrial calendar. All true Christian religion is secular just as its founder. The college calendar, more and more, is the bulwark of free time against factory time. Therefore it is a religious institution of Christianity. You all know scores of cases where the yearning of a man for participating in our meaningful body of time wins out over the fragmented chopped-up times of the business world. That's why people move to college towns. We live one great "Time-bettering Day"; its nine months are like one day. The college calendar represents the Church against the State, is heaven against hell, time against space in our society. It creates through its unreal world of education and matches a body of time comprised out of time bettering days! I am afraid that it may not forever retain this great function. Our time bettering ways are threatened by the military and the economic powers as well.

As a warning, I would like to show you now how once before a heavenly calendar was destroyed. Our academic calendar constitutes a re-conquest of freedom from nature's laws. Heaven has come back in the shape of the secular college year; 8–1/2 months of impersonal, 3–1/2 months of personal life, we now enjoy. But this has evolved after another great way of heavenly life on earth was abandoned. A look on a chart on which all the calendars are shown as they intersect and overlap, speaks loudly. On it, you would find the ecclesiastical calendar. While we all together live the first month to the final weeks of commencement, through the football weekends and exam periods, every one of us in splendid isolation may attend chapel or Mass or Sunday sermons in private. He has Lent from March 3 to April 12. By the way, we are in the midst of Lent tonight. Yet, it verbatim is "by the way" solely, not on our common road. He may try to celebrate the great Week of Week of Weeks, from Palm Sunday to Easter Monday. But he has classes on Maundy Thursday, on Good Friday and on Easter Monday. Nobody more than the Puritans murdered the whole church calendar and emasculated it so as to consist of 52 sabbaths; today this austere colorless Judaising has lost its hold on the community more and more. But their negative success is with us. The canonical hours are forgotten. We are the heirs of a great catastrophe (or a rebirth; we call it the Renaissance) which did away with a previous order of time-bettering days. When I hear the Humanists sneer at the downfall of ecclesiasticism, I often wonder how the humanists can overlook our own frailty. Our academicism is now under the same severe attack against its calendaric privileges.

We soon may be mourners ourselves. Hence a 2nd chart is needed to explain to us the downfall of our predecessors. It would show the canonical laws of monasticism and the working laws of medieval craftsmen, in their contrariety.

Again I may begin with Shakespeare. Shakespeare has two plays whose titles suggest a calendar unknown today. One is *A Midsummer Night's Dream*, the other is *Twelfth Night*. Both titles are remnants of a period when no abstract Eastern standard time, no day of 24 hours, no year of 365 days with a leap day every four years, was contemplated. It was this the period preceding Shakespeare's own times. Shakespeare then has left this period behind and is our contemporary with regard to abstract time. For in the very Midsummer Night Dream, King Theseus already exclaims "the iron tongue of midnight has struck twelve." In Oberon's and Titania's real days, however, Theseus could not have said this. Because down to the death of Dante the day that begins at midnight and ends at midnight was considered an anti-natural day. The Roman lawyers of the ancient Republic had squeezed in such a sacrilegious reckoning from midnight to midnight. And not before the Roman jurists was there such an abstract "civilian" or "civic" day. "Diescivilis" was different from "tres naturalis", "dies fastus", "dies nefastus". The lawyers invented it to get away from the divine days and from the natural days as well. You see the enormity of the civic day from this fact: The Greeks had not even a term for such artificial legalism. In Greek, the oldest occurrence of a term for the 24 hour day is in the Greek text of the New Testament. In "nature", day and night were two separate entities.

But lawyers needed the civil day from midnight to midnight, as we need Eastern standard time, for their legal statutes of limitation, of legitimation, of terms and bans and inheritances. When can it legally be held that people die or act on the same day? These were practical questions and the reckoning from midnight to midnight was a legal fiction, a typical abstract process of reasoning. We don't realize this as you and I have become totally abstract in our relation to time. We get up at a certain hour regardless of sunrise or sunset. This, however, was not done before Shakespeare. That this play, "Midsummer Night's Dream" was dated on the longest day of the year,[5] that we today are having equinox being March 21, then made a difference of which none of us who assembled here tonight, has an adequate experience anymore. The hours of the day, down to 1500, were subdivisions of the night or the day as actually experienced, reaching from the actual

5. It was acted on a very short day, in January!, for the first time.

sunrise to the actual sunset, and their experience of these days or nights meant observance, obedience and submission.

True enough, this daytime or this nighttime was subdivided into 12 hours each. And this fact misleads us when we hear of it. We too, have 12 hours. But the length of these twelve hours could only be 60 minutes on the Equinoxes. Today, the peoples of the world before 1500 and we agree because it is March 21: twelve hours of the equal length of 60 minutes stretch from sunrise to sunset. However, at the two solstices, the twelve hours though still twelve, had a very different content. On a June 21st, sunrise might come at 4 in the morning, sunset at 8 in the evening. In that case, 16 times 60 minutes were to be divided by 12 as there always were to be 12 hours. Hence each hour 80 minutes long. On December 21, with a day of perhaps 8 hours of sunlight, the single hour was composed of 40 minutes only. For the day held twelve hours in analogy of the year's twelve months, and no intent for equal length intervened.

The real movement of sun and stars, and real change of light and darkness forced the perpetual re-ordering of the individual hours. Medieval man did not live abstractly. Modern may believe in the fancy 'natures' of Thoreau, of Rousseau, of Thomas Paine. But these natures are throughout abstractions which we call 'nature' which however, have nothing to do with our personal or political existence or our five senses. It is a scientific concept, not an experienced reality. We moderns command sun and moon by our standardized regional abstract times. We violate experience by our mental tyranny. And we indeed have to boast loudly of our interest in facts. For in fact, we are interested in conventions.

The concretely living people of the Middle Ages felt that the mind should submit to the observed facts. They demanded from reason the inconvenience of changing its abstractions for experiences' sake. Having little artificial sources of illumination, they were forced to concentrate on daylight. Winston Churchill has reminded us of this when in World War I he created our 'summertime'; we by now may recognize this one benefit of a sensually experienced day. But there is one other advantage of bygone calendaring which for us is much more difficult to realize. In fact this advantage is lost on most modern men so totally that they think it absurd. Even at the risk of being charged by you of absurdity, I have to bring this greatest advantage to your attention. The title of Shakespeare's other play Twelfth Night, contains a hint totally forgotten today; it proves that physical, observable, and observed time may speak a language which we have lost to our lasting damage. These Twelve Nights so-called stretched from December 24th to January 6th. However, in some parts of Europe, these 12 nights were duplicated. 12 nights were counted from January 6th in two opposite directions. One dozen ran from the Day of the Three Magi backward into December, and the other dozen ran forward towards the 12th of January. In this doubling up, the profound meaning attached to the formation of this unit of Twelve Nights, is more easily explained than in its reticent rudiments in the title of Shakespeare's play. For, in Southern Italy where the doubling is found, the days carry the eloquent names of months. January 5th, for instance, and January 7th both would be called February, January 1 would be called "June", and January 12 also bears the name of the month of June. "December", figuratively speaking, was used as the honorific title, for January 17th and December 26; the wintry quiet calm of the home reflected on the secrets of all time. The halcyonic days, then, represented the 12 months of the year, in advance and for ritual and liturgy. People impressed on their minds in these days of preparation and rest the mysteries of the whole year. The unit of time, therefore, was a living cell to them and as a living cell contained all the elements of the larger life, these twelve nights were felt to contain the secrets of the whole of time. As a day may represent a month, so, in the Greek Olympiad, the four years really re-presented four times 365 years. In this radical manner, the Greeks mastered the Egyptian Great Year of four times 365 years, *id est* 1460 years. The Egyptians did celebrate this "Egyptiad" so to speak, in 1321 B. C., and in 139 A. D.

The Greek Polis with its minute territory imitated the principles of Egyptian fertility and settlement; hence, it is not to be doubted that a parallel foreshortening, parallel to the foreshortening of the political space, occurred in the Greek Chronology. The four years between the Olympiads meant ritually and liturgically the same as the Egyptian Great Year.[6] In our Church, the time from Christmas to Pentecost makes present the whole life of her Founder, *id est* at least

---

6. Compare [Heinrich] Zimmern [1862–1931], "Der babylonische Volltag ist ein Rundjahr en miniature", *Berichte . . . Leipzig*, [vol.] 53 (1901), 56, and on the workers' day above p* 3.

30 years. And the second half of the ecclesiastical year is intended to make present the whole history of the Church and of our Era, more than 1900 years! In Egypt, one month was used to symbolize 30 years, and one season of 120 days symbolized 100 years; this saeculum, the period of 120 years acquired a separate role with the Etruscans whose total chronology was based on the 120 years. But the number was taken from the tripartition of the year of 360 days in three seasons of 120 days each. This should open our eyes to the organic experience of the times. Unless you admit it into your system, you remain impotent with regard to time. The liberals can neither experience nor master the time process. Small and large units of time could replace each other. Just so, a medieval cathedral led the worshipper through the six ages of the world's history simply by moving him forward from the gates towards the apsis. We, therefore, should not be surprised that the same principle prevailed in ordering the celestial life of the angelical hosts in the monasteries. The canonical hours of one day aptly represented the 7-day week. Prime at sunrise, terce, sixth, none, vesper and compline divided every day of the monk's life into the whole cycle of all times. The sun's day represented the week: the night was his sabbath, with her laudes in the dark. Correspondingly the week had 7 days, the world 7 ages (the seventh lying at its end as the Eternal Sabbath). What the pre-Franklin, the pre-pragmatic mind called "time" was not a quantity, but a melody. Any particle of time could be lived as eternity and that it was lived as eternity was expressed by the 7 parts into which it was subdivided.

Praying 7 times daily also meant to daily live through the Seven Days of the Great Week of Easter, and since the Great Week of Easter is the Queen of all the other 51 weeks, indirectly one single day was lived in the light of all times. One other aspect of this pre-gustation of eternity in the midst of the time stream, deserves attention. The times moved towards their highest point in a crescendo. The whole year of the Church waited for Easter as its climax. The whole year of the old Israel pressed towards the Day of Atonement. The whole year of Egypt, small and great, rushed downhill eagerly from the meaningful five days of the New Year, from which everything else received its significance. In China, in Rome, in the Maya Calendar, one Moment betters all the other moments of the cycle. This New Year Day attracts or lifts up the otherwise disheveled wrecks of time and forms the necklace of diamonds around the neck of the Lord of Time. Since we have lost this immediate power to accept a particle as the representative of a whole, most of the language of Church and Synagogue, of antiquity, is lost on us. It is analogous to modern man's hesitation in speaking of God's finger or womb or eye or smile or of the Church as His bride. Modern man calls such expression embarrassedly "similes" or metaphors. But he may rest assured that his own finger only got its name from its quality of there also being God's finger, and the Day only was observed because it stood for eternity. "Metaphor" so called is the genuine source of all speech and language of our race, and the languages of anatomy and dictionaries are much later layers of language, derived by abstraction from the "Day of Days" and the Back of God. The parts of our body as well as the parts of time have received their names only after and because they proved serviceable to express some divine and eternal meaning. As metaphor for our experience of calendaric times, high and low, our language has been born, and is reborn. Scientific language is rundown, expired, murdered language. If you do not reconcile yourself to this origin of our words in potent speech, you will never understand the order of monastic life or of liturgical living. For the hours in a monastery shifted in their lengths according to the seasons of the year. In other words, on no two days in sequence did the word "hour" signify the same length of time. This, to modern man borders on the insane. Our hour has this one and only merit—that always it is sixty minutes long. This would have been to its disadvantage in antiquity. For the ancients, pious pagans, pious Jews, pious Christians, all were in agreement that man could not abstract time out of the hands of the God who sent each day, each moon, each cycle of the firmament. It would have been blasphemy to teach the times our own minds' lesson that they had to behave according to our abstract scheming. In the continuation of the unanimous reverence of five thousand years, the monk's seven canonical hours observed God's time. Therefore, the hours could not help running from the actual sunrise to the actual sunset as observed "here", let us say, in Monte Cassino, and "now" on October 1, 529 A.D. Only in this way could prayer enter upon God's real times as they ringed the eons of eons.

Both aspects of the pre-Renaissance time sense are lost on us. For instance the King James Version mistranslates the organic flow of the eons by its heretical space concept "world without end". But the Church suggested the organic flow of times after times, epochs of epochs, and nobody today can share the time experience of the apostles unless he casts out the abstract dead time of "world without end". The genuine formula says the very opposite. It says that the end of the world must be experienced at the end of an epoch as much as the beginning of the next. Only by living through the end of one eon first and the beginnings of the new eon first[,] and thanks to this experience, the end of the previous eon, later, can we realize the times. Eons are like links in a chain. We have to hold on to the next eon as it shapes up in catastrophe, and let go the previous, a ring completed in a catastrophe. How else can we realize the Lord of the eons of eons? For God survives the end of time; with the revealing words, "And the end of the world was long ago", [which] begins a great song. God survives ends as well as beginnings. Now exactly this truth we have lost and so we mock and are mocked by the believers in Armageddon. The end of the world, the monks realized every night. Time died. The day died. And then it rose again. For this great existence of the death and resurrection of Christ in their canonical hours, they joyfully paid the price of shortening and lengthening the twelve hours. These twelve hours themselves after all were there in honor of the twelve months during which the great year of atonement, the great New Year's Day was brought back over 360 degrees. In other words, the number twelve, in the hours system, was as much an organic reference to the twelve months as in Shakespeare's play "Twelve Nights". The Twelve were meaningful with reference not to the single nights but to them as they were reminding us of the twelve months. Has not God Himself created the numbers of the heavens? Man obediently depicts them in the shorter units of his daily rhythm. Don't think me too pedantic that I should ask you to re-live this attitude that one day, twelve days, one Olympiad, may give us the revelation of greater structures of time. If you insist, as rationalists, that you want to believe in the unspeakable coarseness of Einstein's time, remember our college year. In it, 9 months constitute one body of time. We should do no violence to compare this body to One Day. Now if nine months may be treated as one day, by us moderns, the ancients, treating one day as eternity did nothing more absurd.

But I have made [had?] strange experiences of modern obtuseness to living time. When a professor of New Testament theology at

Vanderbilt read the proofs of my book *The Christian Future* he got mad at me because in preaching against the Puritans I had written that 52000 Sundays before God were as one Sunday. "That is not true," he verbatim shouted. I quoted, "A thousand years in Thy sight are but as yesterday when it is past, and as a watch in the night." He has never forgiven me this attack on his rational preaching, as though on a Sunday we were just on that one day. I would not mention this man's anger if most humanists did not ascribe to us believers the most atrocious superstitions about time. Take their notion that the Old Testament had the world created on six ordinary days of 24-hours each. The truth is that our earth days have always been treated in religion as a mere image of the Great Days of God's Time. For instance, St. Augustine in his Commentary on Genesis says: "The days from sunrise to sunrise in which we are engulfed we believe to exhibit a vicarious analogy and we must not doubt that ours are not similar to God's days but far below them." (4, 27, 44). All this power of projecting our short-lived moments on the dial of the plenitude is forgotten among the so-called educated people. The humanists are the true barbarians today with regard to that time which for Cartesius, their ancestor, was a daily miracle. It therefore is not a trifle that St. Augustine's power to see in God's Days the model, and in man's small days a poor imitation, once has filled whole nations with the creative power of molding life.

The canonical hours were the monk's way of life, and the monk's way of life between 300 and 1600 was civilized man's way of life. In Spain, in Shakespeare's days, one-third of the population were monks or nuns. Shakespeare himself in "Measure for Measure", makes Queene Anne, King James I's wife, who had become a Roman Catholic, appear as a Poor Clara, *id est*, his heroine Isabell marries the king as a Poor Clara of the Franciscan religion. If a queen of Protestant England in Shakespeare's days and one third of Spain observed the canonical hours, you will more easily envisage the universal domination of the canonical hours over the thoughts of the peoples between Alfred the Great and Thomas More. To identify the two rhythms, that of eternity and that of one day, became a living experience of the multitudes. Now the more the Church looked for the image of eternity within one day's sunrise and sunset the more the earthly part of man rebelled. An image is not identity. Analogy is not sameness. Workers must toil by the sweat of their brow. Gravity claimed its right, as it always will

when the mind's "heaven" usurps too much. This is of universal validity. Science, too, this heaven of abstraction, and its machine age
came to their human frontiers when a physician was able to tell the
British House of Lords that 23 hours of work for the children was
not damaging to their health; so 500 years before this doctor testified
such nonsense, the citizens of Europe protested the nonsense of the
canonical hours for the craftsman's work. Good for heaven, the monastic timetable was harmful to the professional man and the farmer.
They sided with young King Henry VI. Man had to tend flock; he
had to milk his cows. Well, I need not go into all the chores of these
earthbound clouts. They had to work by the sweat of their brow, and
they had to eat and sleep. These necessities are our earthly part and
they defy the skylark's heavenly rhythm.

Already the Founder of the Benedictine order had found out
the discrepancy between earth and heaven. As he demanded, *"ora et
labora"*, both heaven and earth were to be served. Hence his monks
had to be rewarded for their nightly prayers by being given a special
rest in the daytime. The nap, during daytime, is the Benedictine gift
to our race. The "siesta" as it is called in Spain is derived from the
sixth hour of the day, the third canonical hour "sext", and it is a Benedictine institution. Between the sixth and the ninth hour, somewhere
between Sext and None, the monks were allowed to make up for their
heavenly psalmodings during the night. It is at this weakest point of
humanity, the natural privation of sleep, that the system of depicting
the angelic choirs on earth broke down. At this point the hours Sext
and None were set in motion and began to become flexible. St. Bene-
dict declared that *"mediante octava hora"*, midway between Sext and
None, the nap was due. By his legislation, then, the accent was shifted
away from the sixth and the ninth hour. Sext and None were canonical, but half-past eight was sweeter. The Benedictine introduction of
the nap presaged by centuries the later victory of noon in the earthly
sense of the hour for the meal.

The emancipation of the secular community was sealed when
the cities of Petrarch's and Boccaccio's days learned to hang bells in
their guild halls which could ring out the hours mechanically. At first,
these bell towers still observed the actual sunset and in Italy, the oldest secular common timepiece opposing monastic time counted 24
hours from one actual sunset to the next. Mind you, the counting of

this 24-hour day did not begin at midnight. There were two obvious reasons for this: the night's hours cannot be reliably observed by the individual and the family in their habitat. But the boon of having all 24 hours counted from the city's belfry as a public service was tremendous. Secondly, the counting from evening to evening was good Old Testament doctrine. Israel's chronology always counts from evening to evening. This, then, was the beginning of the new chronology. In France, however, this Italian 24-hour city clock was replaced by the 12-hour reckoning. And it seems to have been in France after 1370 that midnight was made the starting point of the new calculus. It may interest you that in Basel to this day the city clock strikes out one at midnight, 2 at one o'clock, and so reminds us of the fact that in the canonical hours, the prime, today as always, signifies the beginning of the whole rank and file of all the hours; and therefore it originally does not signify the completion but the start of the first canonical period. Of this clock in Basel the other Swiss rightly say: "Oh, the people of Basel are behind us a whole century, but ahead of us by one hour." This is literally true, as this canonical count is by a century obsolete. But it seems eternally meaningful to treat midnight as secular man's "Prime".

Elsewhere this last influence of the canonical hours has disappeared. Midnight now is 12, and noon is midday. Heaven's None of prayer (=3:00 p. m.) has become earth's noon of eating. And we live our earthly day now as an abstraction from midnight to midnight. Heaven as well as earth have lost their power over our lives. And man's God is the alarm clock and the railroad time. This is the hour, the Nomos without creator or creation. We live conventionally. I think this story has a moral for the future. The monkish tampering with eternity had placed every day in the full light of the absolute. The world has not tolerated this illusion. If before God, one thousand years are as one day, it does not mean that by us one day can be treated like one thousand years. In other words, ecclesiasticism, more correctly monasticism, ended because history puts the real times of our race, centuries and generations, beyond the small rhythm of any one 24-hour day. The future task of mankind can never more be to secrete the secrets of the eons into 24 hours. If I may be allowed to express my own conviction: decades, generations, time spans of five, of fifteen years will have to be experienced and cultivated like those units of three scores and ten which Lincoln still could quote. The real mystery of time then will [not] be found by analogies between day and millennium, but by patiently "doweling" day and week and year and generation.

I believe in pluralism. We need the intersecting of many rhythms of time. Our stomach and our consciousness respond to a 24-hour rhythm. Our faith and our hopes respond to centuries. Our noble passions like the love of husband and wife, of veterans, of sects, rule time spans of 25, 30, or 40 years. The 24-hour day and the week, the month and the year, should not becloud the spheres of greater revolution. The chronology of family succession, of wars and peaces, has been destroyed by the heresy that the mechanical time clock revealed all there is to be lived in time, by time, and by timing. This mighty republic during 1952 nearly lost its place on the globe because she waited for the election of a new president and did deny "for the time being" the higher sphere in which the U. S. must move and time its actions.[7] And when this absurdity of our Constitution is mentioned, people laugh. They shrug it off as funny. "But whilst this muddy vesture of decay dost grossly close it in, we cannot hear it." Cannot hear what? "There's not the smallest orb which thou beholdest but in his motion like an angel sings . . . such harmony is in immortal souls."[8]

This leads us into the last and fourth part of our question. Obviously, its topic and its content are forced upon us by the outcome of our attempts to establish the days that better time, in the past. There was the calendar of mental progress, of better poems to be written year after year; in the poet's sonnets the quality was going to be perpetually improved. From this faith of genius, sprang the idea of technological, of scientific progress, and it has accelerated time year after year in the end to supersonic speed. There was the college calendar of mental growth, of a body of time to accomplish the birth of the next generation. There was the attempt of improving every day by bringing heaven to earth in the canonical hours. With monasteries and nunneries training everybody in analogical living, as though one

7. *Ed.* It is not clear what Rosenstock-Huessy is referring to here. In March 1952, Pres. Truman announced that he would not seek re-election. At the Democratic convention in July, Adlai Stevenson was named the party candidate for president. The Republican convention in July chose Dwight Eisenhower. Eisenhower was elected president in November, but in accordance with the law did not take office until January. Meanwhile the U. S. was at war in Korea.

8. *Merchant of Venice*, Act. 5, scene 1.

day embodied eternity, the one year of the liturgy embodies the whole life of Christ and His Church. Any monk tries to live this analogy and at one time the laity was persuaded to nearly follow the monks. This monasticizing of the laity had overreached itself by 1300. In the city belfry, in the 24-hour day, in the community's proclamation of its own time, the Franciscan and Dominican utopia of "everybody a monk" exploded. What Boccaccio meant in Italian history, the fixed immobile term "high noon" at 12 o'clock symbolized for England. In it, Earth moved toward its own earthly law. We may call this "man's re-naturalization". And this is the Shakespearian meaning of "nature". It exclusively came to mean that part of creation which is the same every day. This concept of nature was quite new. Shakespeare's term "nature" is not that of the scholastic theologians: the world of cows with two heads, of fairies and of man's fallen nature. No, now it is that part of reality which is below man, which is brought under our laws. In the 24-hour day, earth was restored to its own rank within the spheres. It was now without analogy to heaven. It followed gravity. Thus, we have built up our industrial system—however, it remained to be seen and to be explored where the limitations of this one sphere, the earthly sphere, might have to be drawn.

For the miracle of time was not abolished. It only now should have shifted to other spheres of history and biography and action. This, the wise have done: Goethe, Van Gogh, Blake have lived pregnant time. But their way of life has been blocked by the "nature-idolatry" of the Enlightenment. And so we now have the sphere of the 24 hour and the solar year, and the chamber of commerce seems in control of our timetables. And the attempts of Goethe or Van Gogh are not electrifying the multitudes as they should. And in their [i.e., the chamber of commerce] timetables spaces govern times as in physics. The result is dead time, wasted time, speeded-up time. The fullness of time has disappeared. Mankind on this continent seems paralyzed in their calendaring, We no longer can select new Time-bettering Days with conviction. Modern Christmas and even more Mother's Day, Father's Day, have become business-bettering days. They certainly do not better the times. It takes more than the proclamation of a Conservation Week by the government to conserve the resources of our lifetimes, to create high days that may claim authority over our low days. For this, the time-bettering day must tower high over every day.

A sad example is the fate of Armistice Day. It now is like the trunk of a mighty tree which the Second World War has destroyed in its meaning. For, neither has this Armistice Day of 1918 been allowed to die nor has it honorably been buried nor has it been replaced by any memorial day of this last war. And this World War II has all the features of sub-human horror. First of all, it bears a mere number: Number II instead of a name. Numbered facts cannot be revealed and they cannot be redeemed because numbered things are without the naming word which alone creates any eternal life. Second, World War II has been labeled War of Survival. This again is a British Darwinian formula, which denies the event an address. Events of the jungle cannot be pacified. When you call an event a War of Survival, you cannot complain that it is not followed by a peace treaty. For in a peace treaty, the two languages of the two warring parties must be blended into one. With War of Survival, one side is alone with itself. With "Capitalistic War" the Bolsheviks are alone with their ideology. It takes a spirited enemy to conclude a spirited peace. The peace always must be at least bilingual!

War Number II, War of Survival—these labels are sub-human. A third label is not so much sub-human as it is discouraging. Churchill has called the war "the unnecessary war." This is good negative theology. The Second World War has exposed the childishness or playfulness of the generations of the twenties of this century. The unnecessary destruction—just because the U. S. declined to heed Bill Mitchell and was unable to cable Mr. Hitler in 1938: "we happen to have 10,000 airplanes"—has created an "anti-calendar situation". Time-deteriorating days press on all the people who have remained childish as they were thirty years ago and wait fatalistically for World War III. World War III certainly will not be a war of survival but of going under, not an unnecessary war, but an unarbitrary war; not World War II, but the suicide of the white race. Men have lost control over time. They follow the nuclear physicists into a space-time continuum. Free men have to live in a time bettering discontinuum towering over space. This whole time-deteriorating slope is reflected in the fate of Armistice Day as it hovers between life and death, peace and war, remembrance and forgetfulness.

Again, I turn to Shakespeare for our recovering. Let me end on a cheerful note by placing before you one of his greatest, yet one of his

least recognized creations of a time-bettering day. It is so magnificent-
ly done that I wish to let it go up before you without appending any
further remarks. Shakespeare today must have the last word as he has
had the first. Therefore my comment must precede the quotation. We
all in our lives know of days in which past and future are illuminated
by our own life's sudden concentration to its greatest transparency.
The calendar of a biography shapes itself of unpredictable, unforeseen,
improvised great days. The 24-hour day of the canonical hours—what
is it compared to the real unexpected Day on which Columbus discovered America or on which you realized your destiny? In such unique
moments of personal decision, all this calendar business seems child's
play. But where do we gain the clarity, the insight, the forms to honor
the Columbus Days of our life? Or to celebrate the secular Days of
God's history with man? In real life they must remain unpredictable.
Yet must they be solemnized and honored as Time-bettering Days.
I had to bury my own father and my own father-in-law; no formal
precedent could solve my problems at these occasions. Every love
story is without cliché or should try to be. Of this, Shakespeare has left
a great example. The highest liturgical wonder of the whole Church
calendar is the liturgy of the Saturday between Good Friday and
Easter. Although rarely known, on this Saturday the liturgy is nearly
secular. "Oh, happy guilt, that has found such a redeemer," the priest
sings. Thus the partition between sinner and saint is withdrawn on
this one day. And in this mood of universalism, the night of Easter is
acclaimed as the night of nights. Seven times the night is addressed as
the night in which one miracle after another was to take place. Adam
and Eve reconciled, Israel leaving Egypt, Jews and Gentiles reunited,
and the Old Israel—this is the final climax—at the appointed hour is
at one with the New Israel. On this heritage of the Great Saturday of
Easter, Shakespeare has built. Shakespeare's *Merchant of Venice* opens
the gruesome abyss between Jews and Christians. But in the end, to
Lorenzo and Jessica, to the younger generation that is, as in *The Tem-
pest*, too, graciously is granted the achievement of harmony beyond
the abyss among the older people. Now these two lovers, Lorenzo and
Jessica, dominate the fifth act of *The Merchant of Venice* and I shall
conjure up in your memories the insuperable lines of the music of the
spheres in the sweet moonlight on the hills. This much you all remember well. However, the eloquence of Lorenzo and Jessica on the day of

their own fulfillment, on their "time-besting day", is not unnourished by a greater calendar. Seven times, they praise this night. And these benedictions are masterful translations into secular thought of the great Saturday's liturgy in the week of Easter. Startlingly as such an analogy has hit me, it seems impossible to doubt that the perpetual formula "this night" of the Great Saturday has inspired the perpetual "this night" in the *Merchant of Venice*. The liturgy has not produced an imitation, mark you well, but an inspiration as a great model. The indefatigable Abel Lefranc has proven this point well.[9] The Time-bettering Days of our lives—we may still live them, celebrate them, solemnize them, Shakespeare seems to suggest to us, in the light of the greatest day of all days, "Mark the miracle of time by God's sacrifice complete." And in this light our own tongue will intone its own song, on our own time-bettering Days.

The way Shakespeare's genius has been led by the liturgy will best be savored by any man of taste for himself. The loss incurred by any merely poetical translation from sacred writ also is revealed through a comparison. In the liturgy it is seven times really one and the same night, whereas in its Shakespearian reflext [reflection?] the enumeration of seven nights is arbitrarily built up. It had to be seven as in the liturgical mode, but the secular traditions couldn't really be made coherent as, f. e., Jewish Paschal and Christian Easter which all work one into the other.

And now compare for yourself. Then let the giant "calendar" accompany you in your religious, your poetical, your personal, and your academic life. Like Ezekial's cherub, the calendar shall stay with you as a plural, a heavenly plural, shaped in forms of heaven and earth, of times and places ineluctably, inexorably, but reciprocally explanatory. May you encounter in this spectrum of the times the complete rainbow of your own everlasting temporality! I myself have built my own faith on the saying in Luke in which the Lord institutes the plurality of "more than one calendar". Strange it is that the first free choice offered by Christ was the choice of calendars; and that, just the same, this pluralism is apt to surprise modern, monistic man. If you, in this surprise, doubt if Shakespeare himself was conscious of the processes

---

9. *Ed.* Lefranc (1863–1952), a specialist in French history and literature, especially Rabelais, argued that the 6th Earl of Derby was the true author of Shakespeare's works. Rosenstock-Huessy was intrigued by the authorship question.

between sacred and lyrical calendar which will be placed before you now, read Imogen's outburst: Her lover's sonnets she calls "Scriptures all turn'd to heresy".[10] Shakespeare the Sonneteer knew too well how heavenly and earthly love borrow from each other's calendar incessantly. And this is as it ought to be to the end of the world.

Sabbato Sancto, id est *the Saturday of Easter Week.*

Gaudeat et TELLUS tantis irradiata fulgoribus: et aeterni Regis splendore illustrata, totius orbis se sentiat amisisse caliginem
(of the whole orb Earth may feel to have lost the darkness!

Haec sunt enim Festa Paschalia . . . .

1.   Haec nox est,
     in qua primum patres nostros, filios Israel
     eductos de Aegypto,
     Mare Rubrum sicco vestigo transire fecisti.

2.   Haec igitur nox est,
     quae peccatorum tenebras
     columnae illuminatione purgavit.

3.   Haec nox est,
     quae hodie per mundum universum
     in Christo credentes, a vitiis saeculi et
     caligine peccatorum segregatos, reddit
     grantiae, sociat sanctitati.

4.   Haec nox est,
     in qua destructis vinculis mortis,
     Christus ab inferis victor ascendit.

     O mira circa nos . . . dignatio
     O inestimabilis dilectio caritatis.
     Ut servum redimeres, filium tradidisti.
     O certe necessarium Adae peccatum,
          quod Christi morte deletum est!

10. *Cymbeline* III, 4, 83ff.

O felix culpa, quae talem ac tantum meruit habere Redemptorem!

5.   O vere beata nox,
        quae sola meruit scire tempus et horam, in qua
        Christus ab inferis resurrexit!

6.   Haec nox est,
        de qua scriptum est: et nox sicut dies illuminabitur
        et nox illuminatio mea in deliciis meis.

Lorenzo:
    the moon shines bright
    1. In such a night Troilus and Cressida

Jessica:
    2. In such a night Pyramus and Thisbe

Lorenzo:
    3. In such a night Dido

Jessica:
    4. In such a night Medea

Lorenzo:
    5. In such a night did Jessica

Jessica:
    6. In such a night did young Lorenzo

Lorenzo:
    7. In such a night did pretty Jessica slander her love

Jessica:
    8. I would out-night you, did no body come. . . .

How sweet the moonlight sleeps upon this bank. There's not the smallest orb which thou beholdst but in his motion like an angel sings still quiring to the young-eyed cherubins. Such harmony is in immortal souls; but whilst this muddy vesture of decay doth grossly close it in, we cannot hear it.

The moon shines bright: in such a night as this, when the sweet wind did gently kiss the trees, and they did make no noise—in such a night Troilus methinks mounted the Trojan walls, and sigh'd his soul towards the Grecian tents, where Cressid lay that night.

Jessica In such a night did Thisbe fearfully overtrip the dew, and saw the lion's shadow ere himself, and ran dismay'd away.

Lorenzo In such a night stood Dido with a willow in her hand upon the wild seabanks, and waft her love to come again to Carthage.

Jessica In such a night Medea gather'd the enchanted herbs, That did renew old Aeson.

Lorenzo In such a night did Jessica steal from the wealthy Jew, and with an unthrift love did run from Venice as far as Belmont.

Jessica In such a night did young Lorenzo swear he lov'd her well, stealing her soul with many vows of faith, and ne'er a true one.

Lorenzo In such a night did pretty Jessica like a little shrew, slander her love, and he forgave it her.

Jessica I would out-night you did nobody come. But hark I hear the footing of a man.

Lorenzo Who comes so fast in silence of the night?

How sweet the moonlight sleeps upon this bank! Here will we sit, and let the sounds of music creep in our ears; soft stillness and the night become the touches of sweet harmony. Sit, Jessica. Look, how the floor of heaven is thick inlaid with patines of bright gold; there's not the smallest orb which thou beholdst but in his motion like an angel sings, still quiring to the young-eyed cherubins. Such harmony is in immortal souls; but whilst this muddy vesture of decay doth grossly close it in, we cannot hear it.

Exultet jam Angelica turba coelorum exsultent divina MYSTERIA ET PRO TANTI REGIS victoria tuba insonnet salutaris tellus orbis caliginem amisisse sentiat, terrenis caelestia divina humanis junguntur in novam renata creaturam progenies coelestis emergat, et quos aut sexus in corpore aut aetas discernit in tempore, omnes in unam pariat gratia infantiam.

EPILOGUE

THIS PAPER WAS READ to the Tucker Fellowship which for twenty years has tried to revitalize the message of President Tucker at Dartmouth

College.[11] When we ask ourselves what his message has been, we shall shed light on the findings of this paper. In return, his message now will stand amplified and amended because he did not have to formulate the rhythmical or calendaric experiences of the Liberal Arts College which this paper had to uncover and to describe for the first time.

Tucker started from the premise which he impersonated in his own life. He himself first was a minister and a teacher of theology, and later became the president of a secular college. Hence, he concluded as follows: Colleges at one time have prepared professional men, divines, lawyers, doctors, and their graduate work. Now a new epoch dawns. Now businessmen will have to step into the place of the professions; therefore, it is they who will have to preach the gospel, the universal priesthood will have to become universal; in fact for the first time in history, the good news is entrusted to everyone who leaves college. This message reached every student of Dartmouth through Tucker's famous chapel service.

How far has this message come true? How far is it a nostalgic dream today? Certainly, the businessmen rule America. Certainly, they carry the ball, whether the ball is the gospel or the anti-gospel. But the secular college has given up chapel, and the departments studiously avoid either to influence or to proselytize. But a much deeper influence has been at work and is at work year after year on all our students. They do make the experience of the good life and the more abundant life if they undergo the rhythm of the college year. What neither doctrine nor personality may do, rhythm does. Rhythm sanctified is the introduction to life everlasting. "God is rhythm" a great poet has exclaimed. We have seen that this in a very profound sense is true. A man without more rhythm than the daily schedule of work or of the news over the radio, a group of men without rhythm [that] which is better than the news of the sensational television show, become ungovernable, a mere mob.

The rhythm of the academic year is as genuine a rhythm as the rhythm of the canonical hours. How colorful was the analogical life on everyday of the monastic life, compared to the drab existence of a modern family who can only talk to each other of baseball, wages, and mileage! Though the community has abandoned its liturgical rhythm, the college renders the community an immense service by insisting

---

11. *Ed.* William Jewett Tucker was president of Dartmouth from 1893 to 1909.

on a rhythm all its own. The educational world officers a worldly liturgy by which the worldly mechanism of the process of production is supplemented.

Fifty years ago, in the heydays of Liberalism, President Tucker sensed all this. But the sacred features of rhythmical time and timing were neglected. The Liberals did not know anything about time and about the shape, "the inscape" as Gerard Manley Hopkins has called the gestalt, the structure of our timespans.

Only after Bergson and William James, after Nietzsche and Franz Rosenzweig, has the West begun to wake up to the dangers of a no longer rhythmical, mechanized time. For instance, William James' Pluralism, pooh-poohed by the logicians and metaphysicists, becomes indispensable for our calendaric living. No one calendar may contain or restrain us. The Sabbath must be observed and broken, and both is equally true. This is pluralism incarnate in our calendaric liberty. The space-time continuum of the physicists, on the other hand, is not the homestead of free men.

Whatever the rhythm of a college, it, in any case, is opposed to the factory continuum, and for this one reason alone it deserves our eternal gratitude because it defends our most sacred liberty. There is in this a very Tuckerian notion, for the liberty of breaking and observing the Sabbath is not a political liberty. It is not listed in the Bill of Rights although the authors of this document certainly wanted it to be implied. This liberty will always defy secular definition. It is a sacred, a religious liberty of our heartbeats. Rhythm has a pneumatic quality, and the spirit blows where it listeth. It cannot be pigeonholed in any one hole as it permits us to freely shift from passing cubicles, offices, halls and places on our rhythmical road through weekdays and "time-bettering days". By the abolition of chapel we are compelled to translate President Tucker's message. Whereas he could state the good life in terms of personality and principles, we are forced to state the same truths in terms of rhythm, of time, of the Calendar. If we do this, we shall be able to defend Dartmouth College from becoming an appendix to the world of the armed forces and the world of production. Both these worlds operate under the laws of necessity. Necessity in Society equals gravity in physics. Necessity is upon us from dead matter, from deadly dangers, from famine, war, disintegration. Armies and industries fight these threats on every weekday of the year. We in the colleges, however, also fight. We fight the rigor mortis, the stiff cold hand of mere necessity. We unearth new pathways out and around mere gravity. We wind up the clock, go uphill in the mountains, soar into the stratosphere, look up to the stars, seek out the depth of fearless souls who defy death.

We defend the message of "time-bettering days", of days and years which shed light on all our ordinary days. Let us defend this message, and then our graduates will do honor to President Tucker's prophecy.

THE GENERATIONS OF
THE FAITH

This paper was published initially in the *Hartford Quarterly*, I, no. 3
(Spring 1961). It appeared later in German in *Die Sprache des Menschengeschlechts*, vol. 2 (1964), 276–300.

"HE WAS A MAN of battle and of creative genius, a man who could
tear apart and could build up, a man endowed with as vigorous an
intellect, as lofty a conscience and, above all, as high a courage as the
human race has ever produced."

This has been said of Jean Calvin by a modern unbeliever, by a
humanist. It is sheer nonsense, for the human race never produces
anything. Quite the contrary, cette race maudite of Adam is itself a
miserable product of the earth unless God recreates its members into
stars in his sky. In this sense Calvin's first biographer, Theodor Beza,
answered the question why we should read his book on John Calvin:
because we should deserve to be plunged back into Egyptian darkness, if we ceased to look up to the stars which have led us out of it.
Not as a product of race or earth, but as a star in the sky, as one witness
in the cloud of witnesses, let Jean Calvin speak here today to us from
his translation of our faith, as it dominated the Western World from
1536 to 1564, and let us speak of its retranslation among us in our
own times by the effort of our friend [Ford Lewis] Battles whom we
salute today. For if this book comes to life, its author springs to life.

More than most books and men, Calvin and the author of the "Institutes" are one. A man of slime and clay is transformed into a star of
history by becoming voice; in this manner his voice and his sufferings
become a word of God for one time: each generation needs such a
voice or choir of voices in God's economy of salvation. We could know
this ourselves. But from our scientific cleverness we often suppress
our own experiences. Must not fathers speak to their sons of their
encounter with God?

The book of Genesis was written out of such experience in the
writers' own generations. Samuel, Saul, David, Solomon, these four
generations had become vocal in a grandiose, painful quartet of
voices. They stand before us to this day. So overwhelming was this
revelation to the contemporaries that the whole Bible took the same
shape. From his heart, the author of Genesis knew how God creates
and so he wrote the creation story from experience: in six toledoth,
six generations, the heavens and the earth also were created by God's
Word. Biblical criticism has ignored these empirical origins of the
Pentateuch.[1] But how could I otherwise speak of "the twentieth century" instead of myriads of seconds? Calvin himself tells us the same
truth. He wrote: "rightly does king David put the times of his youth
into the plural. For, without God, there only are incoherent moments
of time." I remain ephemeral as long as I babble myself. The little
devils sell me short. Only God's commands can create units of time,
id est, epochs, ages, generations, centuries, eras. As you know in the
Bible, 'eternity' is not timelessness, but is literally, the recurrence of
epochs.[2] God creates the epochs by our obedience. Calvin thunders:
"*La première règle c'est que nous aions la bouche close et qu'il n'ait que
lui qui parle et que nous ouvrions les aureilles pour escouter. Nous ne
somes que ses organes et ses instruments.*" Only God's word creates that
which we may call our times, our epoch, our age, our century. When
we let God speak and listen and behave as his tools and instruments,
only then can the times coalesce. "*Dieu seul règne et tout le monde soit
assujettie à lui, brief qu'il ni ait que sa parole, qui ait toute audience,*

---

1. Only Benno Jacob (1934) clearly stated the true relation of "Samuel" and "Kings"
to the books of Moses in his masterful commentary to "Genesis."

2. The liturgical phrase "world without end" is a totally unwarranted mistransla-
tion. Vide my *Soziologie* II, *Die Vollzahl der Zeiten* 1958 S. 384ff.

sans que personne y ajoute un seul mot."[3] God alone is king and the whole world is subjected to him, in short, nothing but God's word should be heard, without anybody adding one single word.

Through Jean Calvin's subjection, the word of God was king for one generation: it reigned for the second day of the Protestant Reformation in the form and shape of the "Institutes" of Calvin's personal piety, humbly generalized by him into 'Institutes of the Christian Religion'. Every generation is a word of God, a line of God's great chant and Calvin is the pentameter as Luther is the hexameter in the distich of the Reform. Because Calvin was the second line in God's couplet of the Reformation his book for the king of France was a task like that of St. Luke when he dedicated his book to his Excellency the Lord Theophilus. Luther had been the great occasionalist, the speaker and writer of the *kairos*, of the appointed hour. Luther is a journeyman of the Spirit. Luther came forward with his 95 theses because the salesman of indulgences passed his house in Wittenberg bodily and presently. And Luther remained the man of the hour, of the inspired moment, of the table talks. Aye, did he not boast that the great light of "sola fide", by faith alone, from Habakuk II, had flashed through his mind "auf dem Gang," i.e. in the bathroom? Calvin was required to condense Luther's daily beads of faith into the rosary of his one book. In moving terms, Jean Calvin has bowed to the miracles of God's timing so that Luther's experience of this free and open and surprising economy of salvation was enshrined in the system of Calvin's "Institutes". His humility in this respect places his textbook outside the range of all the other academic or scholastic textbooks. Calvin, most successful systematizer, preacher, textbook writer, reverently placed his own skills in the second line of the couplet, in the second generation of the reformed faith. For he humbly wrote: "not the routines of preaching convert. God's ordinary economy and dispensation by which he calls his own children follow no unvarying rule. He may use other ways. Certainly God has used many another way of giving a man true knowledge of his maker by inward means, that is by some illumination of the Spirit apart from the medium of preaching." In these lines the teacher and unexcelled systematizer, Jean Calvin, voluntarily has taken second seat in the economy of salvation, just as

3. Calvin, 69, 446.

Paul did when he, the greatest of teachers, cried out *"Scio cui credidi."* I know to whom I have given my heart.

The child Jesus was not a prophet, not a teacher or rabbi. For this reason, after Christ, all we professional people have to be given the slip time and again lest God become predictable. We shall not come to the end of this memorial hour before recognising that Calvin's notorious doctrine of predestination represents a parallel reverence before the economy of faith, parallel to the one of which you have just heard here in his refusal of accepting any foreknowable monopoly for preaching or teaching. Mostly the interplay of the successive generations of the spirit is glossed over, as it is glossed over for Peter and Mark, Paul and Luke, the Aramaic and the Greek Matthew. Luther and Calvin are lumped together as the reformers, or we hear that Calvin came a little later than Luther, and it is true that the 'Institutes' were written nineteen years after the 95 theses. Figured mechanically, 'nineteen times one' is not impressive. But it is as with Hegel and Marx. Only 17 years separated Hegel's climax and Marx's Communist Manifesto. Nevertheless Marx lived a whole epoch apart from Hegel. Calvin was separated from Luther and Melanchthon not by nineteen years but by an abyss. The abyss between Hegel and Marx obviously was the proletarian disillusion with the bourgeois ideas, a disillusion of which Hegel had no inkling but which visited Marx. The abyss between Luther and Melanchthon on the one side and Calvin on the other, was opened in the peasants' war of 1525 and the anabaptist movements, reaching its depth in the New Jerusalem of the anabaptists of Münster in 1535. Please, present to your mind this fact: Calvin began to think, to formulate, to write after the potential abuses and limitations of the Reformation herself shone forth. Not from an in itself meaningless external chronology should we call him the authoritative voice of the second generation.

Alas, the humanists think of man's generations in astronomical terms. But in God's economy of history a second generation is required as soon as the utmost, the extreme consequences of the first generation's novelty in action may be assessed. In this sense, for example, Chief Justice John Marshall embodied the second generation for the Common Law in the new United States. In this same sense, Calvin is not free—as Luther had been—to speak out regardless of ranters, of antinomians, of anabaptists and all the proud doctrinarians of the Reformation. Calvin has become the man, the voice, the power of this second generation. As a Lutheran who later became a deacon in the French Reformed Church in Frankfurt and who in the U. S. worships in the Congregational Church for a generation, I had much cause to ponder over the dialectics between Luther and Calvin, and I deem it one of the open desiderata of our Sunday school instruction that this dialectics be used for edification. For, it reveals a perpetual crux of our faith.

Our faith is meaningless unless it receives its doctrine from history. The Bible history is the source for our teaching. Calvin lived immersed in this necessity. And in as far as he did, his book itself in a measure has reached the stature of an inspired creature. This cherub of the Reformation did not dish out classroom generalities. He voiced an emergency in history; hence it should not be labelled 'Institutes of Christian Religion', it is Calvin's account of his own piety. Here, however, we come to the limitations of the man's self-understanding. He did not know and he did not wish to know that his was a place in history. He made himself smaller than he was. He introduced his book with this misleading sentence: "However the knowledge of God and of ourselves may be mutually connected, the order of right teaching requires that we discuss the knowledge of God first, then proceed afterwards to treat the latter." This is the wrong order. It vitiates the whole book, and, by the way, all theology, as it cannot help generalizing God into the God of Aristotle, far away from myself. Soon we shall have to explain Calvin's famous and frightening chapters on predestination as the result of his kowtowing to our inherited and unbiblical order of teaching. They still dare to teach among us the divine mysteries per se, abstracted from your and mine and Adam's and Eve's and John's and Joseph's and Mary's and Luther's and Calvin's encounters with God. But outside these encounters we may know nothing of God. Because Calvin seems to omit them purposely he has to be supplemented today. Fortunately the real fact is that he was bound by history. Calvin was called forth by the historical crisis into which the Reformation had driven. By now you may be more willing to listen to my request of today that we should celebrate Calvin himself as a translator. In your mind the translation of our friend Battles may seem the translation of an original work, the 'Institutes' by John Calvin.

I defy you on this. Today's celebration would not be more than sentimentality unless we trace the history of the spirit as a sequence of translations. Yes, today we do celebrate the *re*-translation of Calvin's celebrated opus magnum. But this man's written word binds together thirty years, one whole generation's Christian life. As God created the generations of heaven and earth in six generations and then he created man, he further created us as generations and he requested us to leave our name, our word of faith on our own time, and from the Bible we may know that under the name above all names every epoch, every generation translates God's word and for doing so comes under the judgment and the name calling and the roll call of our creator. While Luther was aging and bodily failing young Calvin already grew into the name-giver of the second generation by translating Luther's genius into lasting doctrine, a veritable Luke of a veritable Paul. But if this be so, then all spiritual life must be seen as translation. And although we still may distinguish translations of the first and the second degree, it is more urgent to consider both, the "institutiones" of 1560 and the "institutes" of 1960 as translations. Could it be that any future doctrine of the Holy Spirit may have to start with the mystery that we all are required to translate, from the days of Adam to the last judgment? Instead of talking of originals and their translations, it is high time to treat the so-called originals as attempts to translate. Calvin's was the task to retranslate. For the Holy Spirit is the translator from eon to eon.

Hence the 'Institutes' had to draw the line against all over-cleverness and overconclusiveness, against all pure reason. He who translates, remains immersed in the water of faith. Innumerable were Calvin's refusals to think for the fun of thinking. The modern mentality of the quizkid he abhorred and, to appreciate his chastity, please face up to the brutal truth that nothing is destroying the mind in this country so wantonly as the right claimed by every unwashed mouth to spit out questions as irreverently as cherrystones. Children and students are fed on curiosity which in itself is just a worthless itch. Certainly this has not been John Calvin's vice. It is difficult to convey his eloquent silences, his reticence. However, when Faustus Socinus pestered him with questions, Calvin wrote, in 1549: "If you wish to know more, ask someone else. For you shall never succeed in your quest of making me from eagerness to serve you, transgress the boundaries placed on our knowledge by the Lord (XIII, 485)." More than once has his greatest experience remained shrouded in silence, as in his decisions of leaving France, fleeing to Strassbourg, returning to Geneva. A great man of Calvin's stature and suffering has described these secrets of the soul's trembling as Calvin has trembled innumerable times. "When horror gripped him despite his longing to do God's will, then something happened which gave him the one thing yet lacking: the decisive shove, compulsion. That eased the strain. On this miracle, mostly, man remains silent, although perhaps we all may taste it once. But it violates our pride. Man seeks his honour in his free act. However in the midst of the act a moment occurs in which man's courage is deficient simply because he has invested all his courage in the act. Unless at this point the shove of constraint is added and helps the act to be born, it never would see the light of day. But this compulsion arrives. Man has an inborn right to be donated with this compulsion, a right which God acknowledges. All prayer ultimately is a prayer to see one's own free will alleviated by this compulsion. All thanksgiving gives thanks for just this. But the shame which surrounds all prayer is caused by this mysterious interplay of our free choice and God's decisive shove."[4]

It is wise to remember the profound reticence around this mystery of Calvin's own prayers when reading the "Institutes". The book and its author have been much abused because of the doctrine called "horrible" by Calvin himself, the doctrine of predestination. I am stressing the reticence because I hope that you may do justice to Calvin's passion for this doctrine with the help of a few tools which I shall try to offer you now: Calvin knew that a book like the "Institutes" represented only a second voyage, a transformation of tales told, of prayers, and of commands, into teachable abstractions. How small was the weight of such abstract syllogisms in comparison to his daily sorrows and conflicts and perils? How often did he have to enter into the agony of solitude, of powerless ignorance which is the fountainhead of any fully personal prayer, of any encounter of a man's unique soul with the creator, in Calvin's term "for our election". In prayer we have to learn that God is not called Almighty because he created the sun and the heavens and the earth. He is almighty because he can conquer all the mighty powers of sloth, cowardice, routine, vanity,

4. Franz Rosenzweig, *Jehuda Halevy*, 2 ed., 1929, on the poem "Zwang."

pride, tradition, law encroaching on my freedom at this very moment. God is almighty not for his horsepowers but for his triumph over all powers in our tiny frightened heart. This very different almightyness was Calvin's central experience. Hence he knew before becoming a professor that which some professors of theology now apparently will have to learn long after they have studied theology, that the language of prayer is and shall remain the soul's first and fundamental keyboard of speech: the intonations of dread and desire, of endearment and of exorcism, of repulsion and attraction are a linguistic reality. The subjunctives of the passionate heart arc more important and more real than the figures of mathematics and the facts of physics. Our school children all learn the wrong logic. For a complete logic would be the whole life of the logos, of God's dialogue with us, about our many ways through his one creation. That which the schools call logic is a ridiculous rudiment. It is a fourth quarter of God's fullness of speech. Does not the logic taught in the schools of the occident only mention the phrases of the indicative? "2+2=4." "The earth is round". But the first quadrant of the universe of discourse consists of imperatives. Even prayer is preceded by commands given and obeyed: come, go, get up, go to bed, look in your heart and write, *tolle lege*, emigrate, become a doctor, *taisez-vous*! Calvin never tired of commanding silence. God compels in his presence that highest of all praise, silence.

When Norway seceded from Sweden the great and very loquacious poet Björnsterne Björnson wired the new Prime Minister Michelsen "now we all must hang together." He received the reply: "Now let us keep our mouths shut." That is divine logic unknown in our textbooks of logic. Yet the validity of this divine logic is a condition of all worship or prayer. The third divine chapter of logic is that of piety, of grateful remembrance. We remember, we narrate, we tell the stories of God's mercies, of men's follies or of our heroes who embodied God's mercy. That means that, as command or prayer, the tales of history also precede mathematics or science. In his "Institutes" Calvin uses the eloquent and untranslatable phrase "*meminerimus*"—"then we shall have to recall", when he feels that history must be safeguarded against scientific logic. And here you see Calvin's dilemma. Writing after the orgies of the ranters he had to step forward into the field of teaching. Teaching exists only with regard to prehistoric man. And I mean pre-historic. The newcomers, the next generation, the laity, the people, the children by teaching are to be recruited for the army of God's fighters. Teaching has to be logical in the diminished sense of mere logic because the laity is prehistoric, the students are this side of experienced law giving, experienced passion, experienced history, *id est*, of the fullness of the logos of God. Because you, dear listeners, expect me to translate the logos of God's commands to Calvin, of Calvin's passionate prayers, the logic of his painful "life history" into the prehistoric logic of this classroom, after all I too have to speak here in the indicative of timeless reasoning, of abstract truth, of 2+2=4. In real life 2+2 never equals 4. Because we have to make sacrifices for each other not the slightest life process can even start, unless the sentence, 2+2=4 is thrown out of the window.

And now you will be able to do justice to Calvin's task in his "Institutes". Our students learn that Calvin lived from 1509–1564, that he reformed the church of Geneva, that his book still is read and now is retranslated by Ford Lewis Battles. They, under the pressure of our world of mechanics, place this with all their other facts. Even the weather they treat as a merely objective fact. Where I am free to shout "What a beautiful day!" "What a horrible season!" they would like to limit themselves to meteorology and repeat the indifferent indicatives of the weather man "It is zero weather". How can Calvin teach these dead souls? How can you speak to these dreadful brats and quizkids who expect to be stimulated, who talk back at random, who base their pride on their I.Q.s? This was Calvin's dilemma as it is ours. And there, to me, lies the explanation of his doctrine of predestination.

Often, in his book, he may seem to drag it in like a red herring and the mild Philippus Melanchthon, this teacher by nature, omitted the whole doctrine. But Calvin was, as we have seen, a teacher by super-nature, by history, by God's call, to embody in a book of instruction the living experiences of the years 1517 to 1536. By predestination Calvin projects the three other quarters of logic, of command, of desire, of telltale into the fourth quarter of the philosopher's logic. For Calvin in all his cruel manner of letting God give his decrees in unending freedom at least abolished the abstract, timeless laws which we deduce. Predestination restored the hidden, the miraculous, the lifeblood of reality, the trust in God to the world of braintrusters. And their world Calvin dreaded. He dreaded students who would never learn to tremble as he trembled when Farel cursed him, invoking God's presence, unless Calvin became the Reformer of the unruly Canton of Geneva. Calvin wrote for our modern students in the abstract academic style of the indicative: "God is such and such. His church is this and this. His sacraments mean this and this". But he wanted these poor minds of the mere indicative to learn of the true God who blesses and curses, who decrees and demands. And how could he translate into a textbook the styles of God and the soul, the language of commands and the language of prayer? His way out was the double predestination. Impassionately he translated the presence of God into the abstract doctrine of his ever inscrutable sovereignty. Predestination projects prayer and obedience, desire and compulsion into the logic of facts. It is a grandiose transposition from the key of faith and communion into the key of reason. I submit that the doctrine of predestination is a heroic effort of translating man's temporality and so-to-speak non-existence and God's eternal existence into the purely spatial concepts of reason. Loyal to Luther's primacy of faith and cautioned by the Anabaptists' frenzies to teach orderly and rationally, he undertook to place God's unending incalculability into the midst of human reasoning. Let us recognize this doctrine of eternal damnation as his attempt to keep the way open for God's presence, as his replacing the insolent descriptions of a ridiculous "God in general" of the philosophers by the only valid form of speaking of God by invoking him in fear as being my God, our God in this very moment, of our being here before Him. For this unacademic trustiness Calvin may strike us as a lunatic. And in fact, when I told an otherwise intelligent humanist, age 75, that I would have to speak here on Calvin, his send off was: "but he was a lunatic!" Well, to this gentleman Thou, O God, art "an object of praise!"[5]

The style of the indicative and of humanism and of logic is unable to transform our minds from their sinful state into one of a new revelation. Logic cannot repent. Calvin's doctrine of predestination attacks logic in its innermost den of 2+2=4. For God may say: your 2+2 do not equal 4. To my own old congregation in Frankfurt, Calvin wrote this on March 3rd, 1556: "*Vous scavez la règle que nous donne le S. Esprit pour nous réconcilier, c'est que chacun cède et quitte son droit.*" The first thing God says is: "2 rights and 2 properties do not equal 4,

5. More on this lunacy of the gentlemen in the chapter 'vivit Deus' of "Das Geheimnis der Universität," Stuttgart 1958.

for your rights and properties are wrongs in my eyes." Calvin is the great translator of God's freedom and of the soul's faith in God's free new action, the two treasures of the Reformation. Calvin translated them into the doctrinal sobriety of the second generation. Listen to these words in the pithy English of F. L. Battles:

> "Human curiosity renders the discussion of predestination, already somewhat difficult of itself, very confusing and even dangerous. No restraints can hold it back from wandering in forbidden bypaths and thrusting upward to the heights. If allowed it will leave no secret to God that it will not search out and unravel. Since we see so many on all sides rushing into this audacity and impudence, among them certain men not otherwise bad, they should in due season be reminded of the measure of their duty in this regard.
>
> "First then, let them remember that when they inquire into predestination, they are penetrating the sacred precincts of divine wisdom. If anyone with carefree assurance breaks into this place, he will not succeed in satisfying his curiosity and he will enter a labyrinth from which he can find no exit. For it is not right for man unrestrainedly to search out things that the Lord has willed to be hid in himself, and to unfold from eternity itself the sublimest wisdom which he would have us revere but not understand that through this also he should fill us with wonder. He has set forth by his Word the secrets of his will that he has decided to reveal to us. These he decided to reveal so far as he foresaw that they would concern us and benefit us.
>
> "'We have entered the pathway of faith,' says Augustine, 'let us hold steadfastly to it. It leads us to the King's chamber, in which are hid all treasures of knowledge and wisdom. For the Lord Christ himself did not bear grudge against his great and most select disciples when he said: 'I have . . . many things to say to you, but you cannot bear them now.' (John 16, 12) We must walk, we must advance, we must grow, that our hearts may be capable of those things which we cannot yet grasp. But if the last day finds us advancing, there we shall learn what we could not learn here'. If this thought prevails with us that the word of the Lord is the sole way that can lead us in our search for all that it is lawful to hold concerning him, and is the sole light to illumine our vision of all that we should see

of him, it will readily keep and restrain us from all rashness. For we shall know that the moment we exceed the bounds of the word, our course is outside the pathway and in darkness, and that there we must repeatedly wander, slip and stumble. Let this, therefore, first of all be before our eyes: to seek any other knowledge of predestination than what the Word of God discloses is not less insane than if one should purpose to walk in a pathless waste (cf. Job 12:24), or to seek in darkness. And let us not be ashamed to be ignorant of something in this matter, wherein there is a certain learned ignorance. Rather let us willingly refrain from inquiring into a kind of knowledge, the ardent desire for which is both foolish and dangerous, nay, even deadly. But if a wanton curiosity agitates us, we shall always do well to oppose to it this restraining thought: just as too much honey is not good, so for the curious the investigation of glory is not turned into glory (Prov. 25:27. cf. Vg.). For there is good reason for us to be deterred from this insolence which can only plunge us into ruin.

"There are others who, wishing to cure this evil, all but require that every mention of predestination be buried; indeed they teach us to avoid any question of it, as we would a reef. Even though their moderation in this matter is rightly to be praised, because they feel that these mysteries ought to be discussed with great soberness, yet because they descend to too low a level, they make little progress with the human understanding, which does not allow itself to be easily restrained. Therefore, to hold to a proper limit in this regard also, we shall have to turn back to the Word of the Lord, in which we have a sure rule for the understanding. For Scripture is the school of the Holy Spirit, in which, as nothing is omitted that is both necessary and useful to know, so nothing is taught but what is expedient to know. Therefore we must guard against depriving believers of anything disclosed about predestination in Scripture, lest we seem either wickedly to defraud them of the blessing of their God or to accuse and scoff at the Holy Spirit for having published what it is in any way profitable to suppress."

Jean Calvin has redeemed theology from Aristotle by holding on, in the midst of the clutter of mere concepts, to God's living and abhorrent mystery.

Let us then celebrate Mr. Battles' translation and Mr. McNeill's edition by placing Calvin himself among the translators. His "Institutes" have transposed, transported, transferred the treasures of the liturgy, of baptism and communion, into the impossible language of reason by way of this doctrine that God remains free to go on creating new times, new people, a true creatio ex nihilo, in every generation and that his children who fall silent before him, may be called to do his creative will for one more generation. Always must the world of man perish, if we do not generate our generation, the next generation of His word. Always the end of our hectic times is upon us. A generation is a creature to be created by our obedience to the true new and next command of creation. History is the chain of translations of God's word in an unceasing stream of generations *"assujetties sous sa parole"* listening to his command instead of speaking insolently out of their own will and arbitrariness and opening their own mouth as a mob may open it for empty shouts, usually thereby murdering Cinna the poet.

Your own translation, friends, is ennobled by this brotherhood of faithful translators, by your brother Calvin's pious translation. In a few examples let me point out how your new edition participates in the common effort of our own generation.

We find it unpalatable to peddle Calvin's abstract doctrine of eternal damnation. I at least do. But I can afford to do this, because my generation looks through the arrogance of the academic style and the allegedly infallible scientific language. Only children, mathematicians, and semanticists believe today that an indicative is wiser than an imperative or a song. God is not a topic for conversation in his absence. He listens in even when students of divinity dare discuss him as a concept as though he did not harken. The tri-unity of the commanding, the beloved and the recognisable God is irreducible to any "Institutes of Christian Religion". We therefore can do without the doctrine of double predestination as soon as we refrain from the lunacy of translating God's overpowering presence into conceptual indifference. Your new translation is protected by this saving grace of our times. Many of you may know how Karl Barth mutinied from his own Calvinistic background in this question of predestination.

Your new translation, instinctively, contains a parallel to Barth's conversion. I find that your new edition at innumerable places is

translating the term, "God's counsel", *dei consilium*, by the word "plan".
To Calvin himself, however, "plan" was to be expressed by "machi-
nation"! And your own translation also uses "plan" at times for this
rather contemptuous term. In most cases, however, your "plan" translates Calvin's "consilium". Is plan counsel? Is consilium plan? Plan did
not exist in 1556. The term "plan" is of revolutionary origin. Since the
world wars and since the Russian revolution, plan widely differs from
counsel. By "economic planning" a consistency is emphasized which
in Calvin was not suggested by the word, "consilium". "Consilium"
points to a here and now deliberation, intention, conclusion; its accent is on the present. It may, of course, affect past and future, but the
ictus is on the present state of mind. In plan, the startng point is the
last date envisaged. From 1969, for instance, a nine year plan would
work backwards to 1961 and 1960.

Yet, my taking note of this change from counsel to plan is not
a critique. The spirit of our times has inspired you. You have moved
away from the liberal dogma of a heap of separated individual souls,
elect or damned, each one of them faced by an inscrutable judge. Your
translation[,] as a living translation always must do, moves us on into
a more complete understanding of God's providence. Karl Barth discovered that God's predestination received meaning only if the soul
turned away from her splendid isolation. Barth asked: "Who is the
most predestined man?" The answer is with Paul in Romans: "Jesus
Christ our Lord." In fact he is the only fully predestined man ever
to appear between Adam and judgment day. The Godman Jesus being free, being beyond life and death, may recreate the patriarchs in
limbo. Where is eternal damnation since Christ entered hell? Christians daily rewrite history, our belief in the triune God changes the
whole picture of predestination as Christ is much more predestined
than any lame, limited, lukewarm sinner's self. The gates of hell daily
may set free another battalion of hitherto damned souls when our
Lord descends.

In the word "plan" this grandiose unity which Ephesians calls the
economy of the fullness of all the times in the "enanthroposis" of God,
reenters our thinking from an unexpected angle. The present world
revolution pushes the term into the foreground. I am reminded of the
birth of the term "*homousios*" in Niceaea. It was a Neoplatonist word
and the philosophical emperor Constantine, not a Christian himself,

used this non-scriptural term. Similarly "plan" is a non-biblical term. The biblical term is economy. However we have lost this term. For the Latin Church and the theologians have translated economy by "dispensation" and this term dispensation today is anemic. It became especially ambiguous as "dispensation" also means to dispense with, make allowance for an exception or to dispense paper-towels. Worse than this ambiguity of the Latin term was the loss of continuity in the use of the genuine biblical word "economy". Christianity pays dearly today for this loss which is born out by your index: the term "economy" is not in it. Yet the Church has been the first world economist as the letter to the Ephesians points out. Alas, we have ceded our most praiseworthy possession, God's economy of salvation, to Karl Marx. We have lost the true economy first by our pale substituting of "dispensation" for it, and then dropping it altogether.

Marx's economic and historical materialism originated from the same kind of heretical necessity with which our pentecostal sects plague us today. The pentecostal sects are indispensable in righteous punishment for our forgetting the third article on the Holy Spirit of Pentecost. Correspondingly Marxism is an indispensable heresy. For no "ecumenical movements" will save us, as they spring from the purely geographical vision. Christianity never moves in space but it conquers death through new joints in time. When the times are out of joint, Hamlet must put them right in his death. This economy of the generations of souls must supersede the economy of commodities. When life triumphs over death the standard of living may lapse. The economy of salvation alone can overcome the economy of secular revolution. Why do our theologians remain blind to their own loss of their best term?

Our friend Battles' splendid indices do show that Calvin nowhere has quoted the locus of this term, the tenth verse of the first chapter of Ephesians. In part, this is remedied by Battles through his using the term "plan" for consilia. For in God's plan Jesus Christ, the most predestined man of God, can never be omitted from any one single man's relation to the predestinating father of our Lord. By using the word "plan" we are compelled to call into every occult counsel of God the comforting presence of the Name who is above all names. Your name, my despondent friends, is not to stand naked and mute against the Judge. You appear under the mighty name of your firstborn brother and King. My own lifework has centered around the parallel task to overcome the Toynbees, the van Loons, the Spenglers and the Gibbons by a true economy of salvation, a "full count of the times". Christ is the Lord of the eons, according to the ancient word: "*si creatura Dei, merito et dispensatio Dei sumus*."[6] Since we are God's creation, we deserve to consider ourselves part of his plan. Thank you, translator Calvini, for your liberality in this use of the word "plan". You have moved one step onward to the times when Christ becomes transparently all in all and may submit all nations and all the eons which are embodied in the nations, to the father. For the purpose of this submission Christ lends every nation and every soul her unending freedom to advance, to break the prison and the spell of diabolical isolation.

I have jotted down many other sentences in your translation which have made me jump. For instance "trencherman" for "*comestor*" is such a felicitous term. Another example: you ask us to "mount up" to God, your realism made me marvel. Calvin rests fully assured that we at all times may change our level which, for an allegedly rigid predestinarian, presupposes a remarkable faculty of free will. We are, after all, capable of being elevated beyond our own system. I should think that a concordance of this single topic might give us a very important help in our strange sea of troubles which a witty Frenchman has well described: "L'erreur en cette manière est de verser dans l'esprit des systèmes alors même qu'on veut y échapper." The error in this matter makes us remain inside the mentality of systems even when we wish to escape from it.[7] But Calvin allows us at all times a spontaneous ascent to God whence to look down upon the systems of the Aristotelians and Platonists and the logical positivists who would like to have us feed exclusively on the dead quarter of God's full logos. We shall need that free ascent for our plight is enormous.

Our generation is not a first nor a second generation as that of Luther and of Calvin. It is a third generation after two world wars. For this reason it now lives in a third sterile time, in the cold war. In other words, three generations have remained torn, unformed, uninformed, inarticulate. We are three silenced generations and their fragments rightly are called "angry", "beatniks", "lost". This time, therefore, these three silenced generations will have to chant the word of God for our

---

6. Paulus Orosius II, 1, 4.

7. Maurice Delbouille, *Sur la genèse de la Chanson de Roland* [Bruxelles, 1954], 166.

three generations together. I am reminded of the songs of old Tyrtaios. Tyrtaios made the three generations in Sparta, the old, the mature, the young, sing together. In today's ambient we have an encouraging symbol. By your loyalty and devotion and industry long distant times are bound again to our own time and by your translating, you fortify us for our own overdue task. Our Reforming Word obviously has as its very theme the rift between the generations. Why are they paralyzed? Because they no longer seriously speak to each other. You, however, admonish us to coalesce with many more than three generations. The gospel generations and St. Augustine and Luther and Calvin coalesce. God's bliss is on those who make his "Holy, Holy, Holy" ring in such a manner that all the ages seem to become One more and more. This is the promised fullness of the times, the remedy in the economy of salvation. Hence it is my privilege to call this day, in thankfulness to God and you, a true holiday.

BIBLIOGRAPHY

Bade, David. "Respondeo etsi Mutabor: Eugen Rosenstock-Huessy's Semiological Zweistromland." *Culture, Theory and Critique* 56 (April 2015), 87-100.

Bryant, M. Darrol, and Hans R. Huessy, eds. *Eugen Rosenstock-Huessy. Studies in His Life and Thought*. Lewiston, NY: Edwin Mellen, 1986.

Cristaudo, Wayne. "Eugen Rosenstock-Huessy", *The Stanford Encyclopedia of Philosophy* (Fall 2020 Edition), Edward N. Zalta (ed.), URL = <https://plato.stanford.edu/archives/fall2020/entries/rosenstock-huessy/>.

———. *Religion, Redemption, and Revolution: The New Speech Thinking of Franz Rosenzweig and Eugen Rosenstock-Huessy*. Toronto: University of Toronto Press, 2012.

Cristaudo, Wayne, and Frances Huessy, eds. *The Cross and the Star: The Post-Nietzschean Christian and Jewish Thought of Eugen Rosenstock-Huessy and Franz Rosenzweig*. New Castle on Tyne, UK: Cambridge Scholars Press, 2009.

Cristaudo, Wayne, Andreas Leutzsch, and Norman Fiering, eds. Special Issue on Rosenstock-Huessy, *Culture, Theory and Critique* 56 (April 2015).

Fiering, Norman. *Understanding Rosenstock-Huessy*. Eugene, Oregon: Wipf & Stock, 2022.

———. "Recent Publications Relating to the Work of Rosenstock-Huessy, 1973-2013". (https://www.erhfund.org/

Gardner, Clinton C., *Beyond Belief: Discovering Christianity's New Paradigm*. White River Junction, VT: White River Press, 2008.

———. *Letters to the Third Millennium. An Experiment in East-West Communication*. Norwich, VT: Argo, 1981.

Kroesen, Otto. *Planetary Responsibilities: An Ethics of Timing*. Eugene, Oregon: Wipf & Stock, 2014.

Leithart, Peter J. *I Respond, Though I Shall Be Changed: Essays on the Thought of Eugen Rosenstock-Huessy*. West Monroe, LA: Athanasius Press, 2023.

———. "The Social Articulation of Time in Eugen Rosenstock-Huessy." *Modern Theology* 26 (April 2010): 197–219. http://onlinelibrary.wiley.com/doi/10.1111/j.1468- 0025.2009.01594.x/abstract.

Marty, Martin E. *By Way of Response*. Nashville: Abingdon Press, 1981.

———. "A Life of Learning." ACLS Occasional Paper, No. 62.

Morgan, George Allen. *Speech and Society: The Christian Linguistic Social Philosophy of Eugen Rosenstock-Huessy*. Gainesville: University of Florida Press, 1987.

Rosenstock-Huessy, Eugen. *The Christian Future, or The Modern Mind Outrun*. New York: Charles Scribner's Sons, 1946.

_____. *Das Geheimnis der Universität*. Ed. Georg Müller (Stuttgart: Kohlhammer, 1958).

_____. *Der Atem des Geistes*. (Frankfurt: Frankfurter Hefte, 1951).

———. *Die Sprache des Menschengeschlechts*. Heidelberg: Lambert Schneider, 1968. 2 vols.

_____. *Fruit of Our Lips: The Transformation of God's Word into the Speech of Mankind*. Ed. Raymond Huessy. Eugene, Oregon: Wipf and Stock, 2021.

———. *I Am an Impure Thinker*. Norwich, VT: Argo, 1970.

———. *The Multiformity of Man*. Essex, VT: Argo, 2000. Earlier editions appeared in 1936, 1948, 1973.

———. *The Origin of Speech*. Norwich, VT: Argo, 1981.

———. *Out of Revolution: Autobiography of Western Man*. New York: William Morrow, 1938. There are several later paperback reprints.

_____. "Pfingsten und Mission". In Rosenstock-Huessy, *Das Geheimnis der Universität*, ed. Georg Müller. Stuttgart: Kohlhammer, 1958, 236-243

———. *Planetary Service: A Way Into the Third Millennium*. Translated by Mark Huessy and Freya von Moltke. Jericho, VT: Argo, 1978.

———. *Practical Knowledge of the Soul*. Translated by Mark Huessy and Freya von Moltke. Eugene, OR: Wipf & Stock, 2015. This is a revised edition of the work published by Argo Books. Originally published as *Angewandte Seelenkunde: Eine programmatische Übersetzung*, 1924.

———. Recordings of lectures: www.ERHFund.org/online-lecture-library.

———. *Soziologie*. Zwei Bänden. W. Kohlhammer: Stuttgart, 1956-58. A new edition in three volumes, edited by Michael Gormann-Thelen and Ruth Mautner, was published by Thalheimer, 2008-2009. *In the Cross of Reality, Volume 1: The Hegemony of Spaces*. Translated by Jürgen Lawrenz. Edited by Wayne Cristaudo and Frances Huessy. London and New York: Routledge, 2017 is a translation of the first volume of Rosenstock-Huessy's *Soziologie*, 1956.

———. *Speech and Reality*. Norwich, VT: Argo Books, 1970.

_____. "Symblysma oder der Uberschwang der Jesuiten." In Rosenstock-Huessy, *Der Atem des Geistes*. Moers:Brendow, 1991. 277-293.

Stünkel, Knut. "Nations as Times: The National Construction of Political Space in the Planetary History of Eugen Rosenstock-Huessy." In *Transnational Political Spaces: Agents-Structures-Encounters,* edited by Albert, Mathias; Gesa Bluhm; Jan Helmig; Andreas Leutzsch; and Jochen Walter. Frankfurt: Campus Verlag, 2009. (Historische Politikforschung, Bd. 18), 297-317.

———. "Eugen Rosenstock's Early Symblysmatic Experiences. The Sociology of 'Patmos' and 'Die Kreatur'". *Culture, Theory and Critique*, 56 (April 2015), 13-27.

Van der Molen, Lise. *A Guide to the Works of Eugen Rosenstock-Huessy: Chronological Bibliography*. Essex, VT: Argo Books, 1997.

INDEX

www.ingramcontent.com/pod-product-compliance
Lightning Source LLC
Chambersburg PA
CBHW061135220326
41599CB00025B/4245